Zeitschrift für Angewandte Mathematik und Mechanik

ZAMM

Applied Mathematics and Mechanics

Founded by Richard von Mises in 1921

Akademie Verlag

Volume 71 · 1991
Number 7/8

ISSN 0044-2267 ZAMM · Z. angew. Math. Mech., Berlin **71** (1991) 7/8, 205–308

Instructions for Authors

ZAMM publishes original papers (main articles and short communications), the lectures presented at the annual GAMM conferences, and review articles (requested by the Editorial Office of ZAMM). Papers in English (preferably) or German to be submitted for publication should be sent to the address at the bottom of this page (master and one copy), in the case of original papers together with an explicit assurance that they have not been published previously in, or are submitted to, another public journal. Authors are held fully responsible for all information included in their paper and should have one copy available for correction purposes. They get two sets of page proofs, one of them has to be returned within one week. Authors may be charged for extensive corrections in the text; corrections in the figures are not possible. They receive 150 (plenary GAMM lectures), 75 (main articles and short GAMM lectures), or 25 (short communications), respectively, free-of-charge reprints. Additional reprints have to be paid for.

Manuscripts should be type-written (one-sided, double-spaced with wide margins at both sides, each sheet numbered) beginning with the authors' names and the title of the paper. Extra types (like bold-faced letters etc.) should be marked. Please, distinguish 1, *l*, e; v, *ν*, *w*, *ω*; *o*, *O*, 0 (zero); *k*, *K*, *ϰ*; *i*; *ι*; *x*, × (cross), *χ* etc. **Formulas** must be written clearly legible (type-writer if possible, never use pencil). Formulas used in the text have to be numbered at the right margin. Use abbreviations for long terms, "exp" instead of "e" for extensive exponents, bold-faced letters for vectors etc. instead of arrows; avoid hierarchies of indices and exponents if possible. The end of a proof should be marked.

Figures should be submitted on separate sheets, drawn with Indian ink and numbered consecutively. Figure captions, tables, and footnootes (if not avoidable) should be inserted in the final place in the text.

References should be numbered in brackets throughout the text and collected according to these numbers in the appended bibliography (References). Each reference should contain Author(s), title, and complete bibliographical coordinates following the subsequent examples.

Books (Reports, Preprints etc.):
IRONS, B. M.; SHRIVE, N. G.: Numerical methods in engineering and applied science – Numbers are fun. Ellis Horwood Series in Mathematics and Its Applications, Ellis Horwood Ltd., Chichester 1987.

Articles in periodicals:
KASSEM, S. E.: Wave source potentials for two superposed fluids, each of finite depth. Indian J. Pure Appl. Math. **18** (1987) 2, 186–193.

Articles in Proceedings etc.:
LABISCH, F. K.: Some remarks on the morphology of nonunique solutions in nonlinear elastostatics. In: KÜPPER, T.; SEYDEL, R.; TROGER, H. (eds.): Bifurcation: Analysis, algorithms, applications. Proc. Conf. at the Univ. of Dortmund, August 18–22, 1986. Internat. Ser. of Numer. Mathematics, Vol. 79, Birkhäuser Verlag, Basel–Boston–Stuttgart 1987, pp. 177–184.

To each original paper and review article, a **summary** of 5 to 10 lines in English (for manuscripts of more than 10 pages also in German) and a short list of key-words and corresponding AMS (MOS) classification numbers are needed.

The precise address of each author with full name, academic title etc. is to be added. Information on address changes is appreciated.

ZAMM
Editorial Office
Rudower Chaussee 5
O-1199 Berlin
Germany

Hinweise für die Autoren

ZAMM veröffentlicht Originalarbeiten (Hauptaufsätze und Kleine Mitteilungen), die Vorträge der jährlichen GAMM-Tagungen und Übersichtsartikel (die von der Redaktion erbeten werden). Arbeiten in Englisch (vorzugsweise) oder Deutsch, die zur Publikation eingereicht werden, sind an die unten stehende Adresse zu senden (Original und eine Kopie). Dem Manuskript von Originalarbeiten ist eine Erklärung beizufügen, daß die Arbeit bisher in keiner anderen Zeitschrift veröffentlicht worden ist und auch kein Antrag auf Veröffentlichung in einer solchen Zeitschrift läuft. Die Autoren sind für den Inhalt ihrer Arbeiten voll verantwortlich und sollten für Korrekturzwecke eine weitere Kopie besitzen. Sie erhalten die Korrekturfahnen in zweifacher Ausfertigung; eine Korrektur ist innerhalb einer Woche zurückzusenden. Die Kosten für umfangreiche nachträgliche Korrekturen können dem Autor berechnet werden. Korrekturen in den Abbildungen sind nicht möglich. Die Autoren erhalten kostenlos 150 Sonderdrucke für die Hauptvorträge der GAMM-Tagungen, 75 für Hauptaufsätze, Übersichtsartikel und die GAMM-Kurzvorträge und 25 für Kleine Mitteilungen Zusätzliche Sonderdrucke sind gegen Berechnung erhältlich.

Die **Manuskripte** müssen mit Maschine geschrieben sein (einseitig, zweizeiliger Zeilenabstand mit breitem Rand auf beiden Seiten, jede Seite numeriert), wobei mit dem Namen der Autoren begonnen werden soll und danach der Titel folgt. Besondere Typen (z. B. halbfett) sind zu markieren. Es soll gut unterschieden werden zwischen 1, *l*, e; v, *ν*, *w*, *ω*; *o*, *O*, 0 (Null); *k*, *K*, *ϰ*; *i*; *ι*; *x*, × (Kreuz), *χ*; usw. Die **Formeln** müssen deutlich lesbar sein (möglichst maschinenschriftlich, keine Verwendung von Bleistift). Im Text verwendete Formeln sollten am rechten Rand durchnumeriert werden. Für lange Terme sind Abkürzungen zu verwenden, „exp" anstelle von „e" für große Ausdrücke im Exponenten, halbfett für Vektoren anstelle von darübergesetzten Pfeilen – Mehrfach-Indizes wie auch Exponenten möglichst vermeiden. Das Ende eines Beweises sollte gekennzeichnet werden.

Abbildungen sind auf gesonderten Blättern einzureichen und sollten mit Tusche gezeichnet und durchnumeriert sein. Die Abbildungsunterschriften, eventuelle Tabellen und (falls nicht vermeidbar) Fußnoten sind an der Stelle einzufügen, an der sie im Text erscheinen sollen.

Literaturzitate sind durch den gesamten Text durchzunumerieren (in eckigen Klammern) und in der entsprechenden Reihenfolge im am Schluß folgenden Literaturverzeichnis aufzulisten. Jedes Zitat sollte Autor(en), Titel und die vollständigen bibliografischen Koordinaten enthalten, entsprechend den folgenden Beispielen.

Bücher (Reports, Preprints usw.):
IRONS, B. M.; SHRIVE, N. G.: Numerical methods in engineering and applied science – Numbers are fun. Ellis Horwood Series in Mathematics and Its Applications. Ellis Horwood Ltd., Chichester 1987.

Zeitschriftenartikel:
KASSEM, S. E.: Wave source potentials for two superposed fluids, each of finite depth. Indian J. Pure Appl. Math. **18** (1987) 2, 186–193.

Arbeiten in Proceedings usw.:
LABISCH, F. K.: Some remarks on the morphology of nonunique solutions in nonlinear elastostatics. In: KÜPPER, T.; SEYDEL, R.; TROGER, H. (eds.): Bifurcation: Analysis, algorithms, applications. Proc. Conf. at the Univ. of Dortmund, August 18–22, 1986. Internat. Ser. of Numer. Mathematics, Vol. 79, Birkhäuser Verlag, Basel–Boston–Stuttgart 1987, pp. 177–184.

Für jede Originalarbeit und jeden Übersichtsartikel wird eine englische **Zusammenfassung** (bei Manuskripten von mehr als 10 Seiten auch eine in Deutsch) von 5 bis 10 Zeilen Länge sowie eine kurze Liste von Stichworten und den entsprechenden AMS (MOS)-Klassifikationsmarken benötigt.

Die genaue Adresse jedes Autors mit vollständigem Namen, akademischen Titeln usw. muß beigefügt werden. Bei Adressenänderung wird eine umgehende Information erwartet.

Zeitschrift für Angewandte Mathematik und Mechanik

ZAMM

Applied Mathematics and Mechanics

Founded by Richard von Mises in 1921

THEORIA CUM PRAXI

Akademie Verlag

Volume 71 · 1991
Number 7/8

The aim and scope of ZAMM is, in agreement with the intentions of its founder Richard von Mises, one of the outstanding scientists of our century in both Mathematics and Mechanics: to publish new results and review articles, the proceedings of the annual GAMM conferences, book reviews and information on applied mathematics (mainly on numerical mathematics and various parts and applications of analysis, in particular numerical aspects of differential and integral equations), on the whole field of theoretical and applied mechanics (solid mechanics, fluid mechanics, thermodynamics) and on mathematical physics.
The journal is of interest for persons working in applied mathematics and mechanics as well as on mathematical or mechanical questions in other sciences, for instance mechanical and civil engineering, electrotechniques, and chemistry.

In Übereinstimmung mit den Zielstellungen des Begründers der ZAMM, Richard von Mises, eines der bedeutendsten Wissenschaftler unseres Jahrhunderts auf dem Gebiet der Mathematik und Mechanik, publiziert die Zeitschrift neue Ergebnisse und Übersichtsartikel, die Vorträge auf den jährlichen GAMM-Tagungen, Buchbesprechungen und Informationen über Angewandte Mathematik (hauptsächlich über numerische Mathematik und verschiedene Teile sowie Anwendungen der Analysis, insbesondere numerische Aspekte bei Differential- und Integralgleichungen), über das gesamte Gebiet der theoretischen und angewandten Mechanik (Festkörpermechanik, Strömungsmechanik, Thermodynamik) und über mathematische Physik.
Die Zeitschrift wendet sich an Vertreter der angewandten Mathematik und Mechanik wie auch an Wissenschaftler und Ingenieure, die sich mit mathematischen oder theoretisch-mechanischen Fragen auf anderen Gebieten, z. B. Maschinenbau und Technik, Elektrotechnik und Chemie, befassen.

How to order

Please place your order with

Kunst und Wissen, Erich Bieber GmbH, a Faxon Company, Wilhelmstr. 4, Postfach 10 28 44, W-7000 Stuttgart 10, Federal Republic of Germany; telephone: 21 07 70, telex: 721 929, telefax: 24 74 39;

or with Akademie Verlag GmbH, Leipziger Straße 3–4, Postfach 12 33, O-1086 Berlin, Federal Republic of Germany; telephone: 2 23 63 50 or 2 23 63 51, telex: 114 420;

or with your bookshop.

Bezugsmöglichkeiten

Bitte richten Sie Ihre Bestellungen an

Kunst und Wissen, Erich Bieber GmbH, ein Unternehmen der Faxon-Gruppe, Wilhelmstr. 4, Postfach 102844, W-7000 Stuttgart 10, Bundesrepublik Deutschland; Telefon: 21 07 70, Telex: 721 929, Telefax: 24 74 39;

oder an die Akademie Verlag GmbH, Leipziger Straße 3–4, Postfach 1233, O-1086 Berlin, Bundesrepublik Deutschland; Telefon: 2 23 63 50 oder 2 23 63 51, Telex: 114 420;

oder an Ihre Buchhandlung.

ZEITSCHRIFT FÜR ANGEWANDTE MATHEMATIK UND MECHANIK

Herausgeber: Institut für Mechanik, Chemnitz. Mit der Herausgabe beauftragt und Chefredakteur: Prof. Dr. Günter Schmidt, Rudower Chaussee 5, O-1199 Berlin, Fernruf: 6745611 oder 6743668 (gleichzeitig Anschrift der Redaktion).
Verlag: Akademie Verlag GmbH, Leipziger Straße 3–4, O-1086 Berlin; Geschäftsführer: Dr. Gerd Giesler, Dr. Bernhard Tesche; Fernruf: 2236350 oder 2236351; Telex-Nr.: 114420; Bank: Dresdner Bank Weinheim, Kto.-Nr.: 754 624 500, BLZ 670 800 50.
Gesamtherstellung: Druckhaus „Thomas Müntzer" GmbH, O-5820 Bad Langensalza, Germany.
Erscheinungsweise: Die Zeitschrift für Angewandte Mathematik und Mechanik erscheint monatlich. Die 12 Hefte eines Jahres einschließlich Tagungshefte bilden einen Band. Bezugspreis je Band 600,– DM zuzüglich Versandspesen. Bezugspreis je Heft 50,– DM.
Bestellnummer dieses Heftes: 1009/71/7/8.

AN (EDV) 35937

ZAMM · Z. angew. Math. Mech. 71 (1991) 7/8, 207–221 Akademie Verlag

ANSORGE, R.; LEI, J.

The Convergence of Discretization Methods if Applied to Weakly Formulated Problems. Theory and Examples

Dedicated to the memory of LOTHAR COLLATZ

Die von STUMMEL *und* GRIGORIEFF *begründete und später von zahlreichen Autoren auf nichtlineare Probleme erweiterte allgemeine Theorie der Konvergenz von Diskretisierungsalgorithmen genügt nicht vollständig der numerischen Behandlung linearer oder nichtlinearer, schwach formulierter Probleme. Eine derartige Erweiterung wird hier zusammen mit Beispielen für nichtlineare elliptische und hyperbolische partielle Differentialgleichungen angegeben. Oft führt der Übergang von klassischen zu schwachen Lösungen zu einem Verlust der Eindeutigkeit, so daß Eindeutigkeit nur durch zusätzliche Bedingungen wie etwa geeignete Ungleichungsrestriktionen garantiert werden kann. Wie solche Restriktionen im allgemeinen diskretisiert werden müssen, um eine Konvergenz der numerischen Lösung gegen die entsprechende eindeutige schwache Lösung zu erreichen, wird ebenfalls diskutiert, und die Ergebnisse werden insbesondere mit den Entropie-Bedingungen von Erhaltungssätzen verglichen. Die Beispiele werden nur behandelt, um die allgemeine Theorie zu motivieren und deren Anwendbarkeit zu demonstrieren.*

The general theory of the convergence of discretization algorithms founded by STUMMEL *and* GRIGORIEFF *and later extended to nonlinear problems by numerous authors does not completely fit the numerical treatment of linear or nonlinear weakly formulated problems. Such an extension is presented together with examples of nonlinear elliptic and hyperbolic* PDEs. *Often the transition from classsical to weak solutions leads to a loss of uniqueness such that uniqueness can only be guaranteed by additional conditions, e.g. suitable inequality constraints. How such constraints must generally be discretized in order to make the numerical solution converge to the corresponding unique weak solution is also discussed, and the results are particularly compared with entropy conditions of conservation laws. The examples are only treated in order to motivate the general theory and to demonstrate its applicability.*

Основанная СТУММЕЛЕМ *и* ГРИГОРИЕВОМ *и расширанная позже многочисленнными авторами на случай нелинейных задач общая теория сходимости алгоритмов дускретизации несовершенно соответствует численной обработке линейных или слабо формулированных нелинейных задач. Дается такое расширение вместе с примерами для нелинейных эллиптических и гиперболических уравнений в частных производных. Часто переход от классических на слабые решения ведет на потерю единственности так, что можно обеспечивать единственность только при дополнительных условиях, на пример при подходящих условиях в виде неравенств. Изучается и вопрос, как вообще такие условия надо дискретизировать чтобы обеспечивать сходимость численного решения к соответствующему единственному слабому решению и сравняются результаты в особенности и условиями энтропии законов сохранения. Примеры только изучаются в качестве мотивации общей теории и показательства его применимости.*

MSC (1980): 65JO5, 65M10, 65N30

1. Introduction

A general theory of the convergence of discretization methods, e.g. discretizations of ODEs, PDEs, integral equations etc. was established between 1970 and 1975 by STUMMEL [23] and GRIGORIEFF [11]. This theory — nowadays sometimes called *functional analysis of discretizations* — was later extended by several authors, e.g. VAINIKKO [24], REINHARDT [19], ANSELONE and ANSORGE [2], [3], ANSELONE and LEI [4] etc.

But all these extensions and generalizations do not completely apply to numerical procedures for the approximation of weak solutions, i.e. of solutions of weakly formulated problems. Hence, projection methods for the numerical calculation of weak solutions of linear or nonlinear elliptic boundary value problems do not fit the general theory. An example will be given in section 2 of this paper. Another example is the approximation of weak solutions of hyperbolic conservation laws by means of finite difference methods as presented in section 3. This hyperbolic problem is at the same time an example for the fact that transition from the concept of classical solutions to a concept of weak solutions can lead to a loss of uniqueness such that the given problem must be completed by additional conditions which will only be fulfilled by the expected weak solution, e.g. by the physically relevant solution. In case of first order hyperbolic conservation law systems, such additional conditions were introduced by LAX [15] and OLEINIK [18], namely certain inequality constraints called entropy conditions because of their physical meaning in gas dynamics. Examples of entropy conditions will also be discussed in section 3.

Also in our general frame, such inequality constraints will later be called *entropy conditions*, and the expected weak solution will be called *entropy solution*.

The question arises, how such entropy conditions must be discretized and added to the numerical procedure in order to single out such numerical solutions of the method which converge to the entropy solution and not to another weak solution of the original problem, or — if there is only one approximate solution for fixed step sizes — to ensure that these solutions converge to the entropy solution when the step sizes tend to zero. The numerical solutions singled out are called *discrete entropy solutions*. We do not necessarily expect that the discrete entropy solutions are unique but we assume that there is at least one for arbitrary fixed step sizes, and we assume that there is at most one entropy solution of the original problem.

The extension of the general theory of convergence to weakly formulated problems will be presented in section 4 and includes principles for the construction of discrete entropy conditions.

Section 5 contains the application of the new theory to the elliptic example of section 2 and improves results from [5]. The application to conservation laws with entropy condition introduced in section 3 is shown in section 6 and yields as a by-result the fact that the Engquist-Osher approximation [10] does not only fulfill a discrete entropy condition locally as shown by STRACK [22] but even globally.

2. A nonlinear elliptic example

Let us consider the boundary value problem

$$-\Delta u = g(x, u)\,, \qquad x \in \Omega\,; \qquad u = 0\,, \qquad x \in \partial\Omega\,, \tag{1}$$

where $\Omega \subset \mathbb{R}^m$ is a bounded smooth domain. Assume the following hypotheses to be fulfilled:

H1) $g(x, u)$ satisfies a Caratheodory condition, i.e. g is continuous with respect to u for almost all $x \in \Omega$ and measurable with respect to x for $u \in (-\infty, \infty)$.

H2) There are constants c_1, c_2 such that

$$|g(x, u)| \leqq c_1 + c_2 |u|^s$$

where $1 < s < (m + 2)/(m - 2)$ for $m > 2$ and $1 < s < \infty$ for $m = 2$.

H3) $g(x, u) = o(|u|)$ for $|u| \to 0$.

H4) $\displaystyle\lim_{\substack{u\to\infty\\(\text{or } u\to-\infty)}} \frac{g(x, u)}{u} = +\infty\,.$

H5) There is a constant $\mu > 2$ such that

$$\mu G(x, u) \leqq u g(x, u)\,, \quad \text{where} \quad G(x, u) = \int_0^u g(x, y)\, dy\,.$$

Problems of this type occur in various fields of chemistry and physics. The existence of classical smooth solutions can only be expected if g is sufficiently smooth with respect to (x, u).

The existence of nonzero weak solutions in the Sobolev space $H_0^1(\Omega)$ with inner product $(u, v)_1 := \int_\Omega \nabla u\, \nabla v\, dx$ of a problem with somewhat analogous hypotheses was proved by AMBROSETTI and RABINOWITZ [1] by means of the Mountain Pass Lemma.

The weak formulation of (1) reads as follows:

Find $u \in H_0^1(\Omega) =: X$ such that

$$A(\Phi)\, u := \int_\Omega (\nabla u \cdot \nabla \Phi - g(x, u) \cdot \Phi)\, dx = 0 \quad \text{for all} \quad \Phi \in J := H_0^1(\Omega)\,. \tag{2}$$

We call J the *set of test functions* or *test elements*. (In this example, $J = X$. This will be not so in other examples, e.g. in Section 3.)

The integral in (2) exists such that the definition of the set $\{A(\Phi)\}$ of operators is reasonable. In fact, from H1) and H2), from the Imbedding Theorem and from

$$\left|\int_\Omega g(x, u)\, \Phi(x)\, dx\right| \leqq \|g(\cdot, u)\|_{L_p}\, \|\Phi\|_{L_q} \leqq c\, \|g(\cdot, u)\|_{L_p}\, \|\Phi\|_1 \quad \text{for all} \quad \Phi, u \in H_0^1(\Omega) \tag{3}$$

we know that the mapping $u \to g(\cdot, u)$ defines a bounded continuous operator $H_0^1(\Omega) \to L_p(\Omega)$ (cf. [14], [17]). Herewith,

$$q = 2m/(m - 2)\,, \qquad p = 2m/(m + 2) \quad \text{for} \quad m > 2\,;$$

$$q \in (1, \infty)\,, \qquad p = 1 + \varepsilon\,, \qquad \varepsilon > 0\,, \qquad \frac{1}{p} + \frac{1}{q} + 1 \quad \text{for} \quad m = 2\,.$$

Now, an operator $K : H_0^1(\Omega) \to H_0^1(\Omega)$ can be defined by the Riesz Theorem:

$$(Ku, \Phi)_1 := \int_\Omega g(x, u)\, \Phi(x)\, dx\,, \qquad \Phi \in H_0^1(\Omega)\,, \tag{4}$$

and this operator is compact. With $A := I + K$, (2) can be written as

$$A(\Phi)\, u = (Au, \Phi)_1\,. \tag{5}$$

In order to solve (2) numerically, let $\{X_n\}$ with $X_n \subset X_{n+1}$, $\overline{\bigcup_{n\in\mathbb{N}} X_n} = X$, be a sequence of finite dimensional subspaces of $X = H_0^1(\Omega)$ and let $P_n : X \to X_n$ be orthogonal projective operators, $n \in \mathbb{N}$. Define approximate operators $A_n(\Phi)$ for $A(\Phi)$ by

$$A_n(\Phi)\, u_n := (P_n A u_n, \Phi)_1 \quad \text{for all} \quad u_n \in X_n\,, \tag{6}$$

such that an approximate problem is now defined by

$$A_n(\Phi)\, u_n = 0 \quad \text{for all} \quad \Phi \in J\,. \tag{7}$$

(7) can not yet be processed by computers, but because of $P_n^* = P_n$, (7) can be rewritten as $(Au_n, P_n\Phi)_1 = 0$ for all $\Phi \in J$, i.e.

$$\int_\Omega (\nabla u_n \nabla \Phi^{(n)} - g(x, u_n)\, \Phi^{(n)})\, dx = 0 \quad \text{for all} \quad \Phi^{(n)} \in X_n\,, \tag{8}$$

and, hence, as

$$(Au_n, \Phi_i)_1 = 0\,, \qquad i = 1, \ldots, n\,, \tag{9}$$

where

$$X_n = \langle \Phi_1, \Phi_2, \ldots, \Phi_n \rangle \quad \text{is given.}$$

(9) now defines a method which can numerically be processed and which can be written as

$$\hat{A}_n u_n = 0\,, \qquad n = 1, 2, \ldots, \tag{10}$$

with operators $\hat{A}_n : X_n \to \mathbb{R}$ independent of test functions $\Phi \in J$. Obviously, every solution of (10) is also a solution of (7) (and vice versa).

If

$$S = \{u \in X \mid u \text{ solves (2)}\}$$

is the set of weak solutions of (1), i.e. solutions of (2), and if

$$S_n = \{u_n \in X_n \subset X \mid u_n \text{ solves (10)}\}$$

is the set of approximate solutions for fixed n, we ask for conditions as weak as possible under which the relation

$$S_n \to S \quad \text{for} \quad n \to \infty \tag{11}$$

holds in a reasonable sense of set-convergence defined in section 4.

By the way, if $S_n \neq \emptyset$ $(n = 1, 2, \ldots)$ and if (11) holds, obviously $S \neq \emptyset$ such that the convergence proof for the numerical procedure can also be considered as a numerically constructive proof of an Ambrosetti-Rabinowitz-type existence theorem for weak solutions of the original problem.

3. A nonlinear hyperbolic example

Let $x \in \mathbb{R}$ be a space variable and $t \geqq 0$ the time variable, and denote by

$$\Omega = \{(x, t) \mid x \in \mathbb{R}, t \geqq 0\}$$

the upper half x-t-plane. We ask for a solution $u : \Omega \to \mathbb{R}$ of the scalar initial value problem

$$u_t + \frac{\partial}{\partial x} q(u) = 0\,, \qquad t > 0\,, \qquad u(x, 0) = u_0(x)\,, \qquad u_0 \in L_\infty(\mathbb{R})\,, \tag{12}$$

where the flux $q \in C^1(\mathbb{R})$ is assumed to be strictly convex, to have the minimum

$$q(0) = 0 \tag{13}$$

at $u = 0$, and to fulfill

$$|q'|_\infty^* := \max\,\{|q'(u)|, |u| \leqq \|u_0\|_{L_\infty}\} < \infty\,. \tag{14}$$

Because of its conservation law form, the differential equation in (12) is often discussed in the literature as a test equation for systems of conservation laws as they occur in mathematical models of physical phenomena, e.g. in fluid dynamics of inviscid compressible flows. (But there are also examples where (12) is directly used as a model of real world processes, e.g. in traffic flow theory [6]).

If $u_0 : \mathbb{R} \to \mathbb{R}$ is sufficiently smooth, a smooth solution exists uniquely in the neighbourhood of the x-axis. But if we are interested in global solutions, shocks can occur even if u_0 is smooth. Thus, only weak solutions make sense in the global situation. The weak formulation of (12) reads as follows:

Find $u \in L_1^{\mathrm{loc}}(\Omega) =: X$ such that

$$A(\Phi)\, u := -\iint_\Omega \{u \cdot \Phi_t + q(u)\, \Phi_x\}\, \mathrm{d}x\, \mathrm{d}t - \int_{\mathbb{R}} u_0(x)\, \Phi(x, 0)\, \mathrm{d}x = 0 \quad \text{for all} \quad \Phi \in C_0^1(\Omega) =: J\,, \tag{15}$$

where $C_0^1(\Omega)$ is the set of test functions continuously differentiable on Ω with compact support.

The integrals in (15) exist for $u \in X$ with $u_0 \in L_\infty(\mathbb{R})$. *Weak solutions* $u(\cdot, t)$ are the elements of

$$S = \{u \in X \mid u \text{ solves } (15)\}\,.$$

We do not a-priorily assume that there are weak solutions. On the other hand, there are simple examples with more than one weak solution:

The Riemann problem described by

$$u_0(x) = \begin{cases} 0 & x \leqq 0 \\ 1\,, & x > 0\,, \end{cases}$$

for Burgers' equation, i.e. $q(u) = u^2/2$, has the weak solutions

$$u(x, t) = \begin{cases} 0\,, & x \leqq t/2 \\ 1\,, & x > t/2 \end{cases} \quad \text{and} \quad \tilde{u}(x, t) = \begin{cases} 0\,, & x \leqq 0 \\ x/t\,, & 0 < x < t \\ 1\,, & x \geqq t \end{cases}$$

(cf. e.g. [21], p. 250).

Different types of entropy conditions have been used in the literature in order to complete (15) in such a way that only the physically true solution fulfills this condition. A rather general description is the following one:

Let $V = V(u, c)$, $\mathbb{R} \times \mathbb{R} \to \mathbb{R}$, be a one-parameter family of functions, convex with respect to u for every fixed value of the parameter $c \in \mathbb{R}$. Moreover, assume V to be continuous and piecewise continuously differentiable with respect to u. Let $F = F(u, c)$, $\mathbb{R} \times \mathbb{R} \to \mathbb{R}$, be another function such that

$$V_t(u(x, t), c) + \frac{\partial}{\partial x} F(u(x, t), c) = 0 \tag{16}$$

holds for every fixed $c \in \mathbb{R}$ and on every subset of Ω where u is a smooth solution of (15), hence even a solution of (12). *V is called entropy function* and F is called the *entropy flux*. If — moreover — $u \in S$ fulfills the inequality

$$V_t(u, c) + \frac{\partial}{\partial x} F(u, c) \leqq 0 \quad \text{for all} \quad c \in \mathbb{R} \tag{17}$$

weakly, i.e. if the weak solution u fulfills

$$-\iint_\Omega \{\Phi_t(x, t)\, V(u(x, t), c) + \Phi_x(x, t)\, F(u(x, t), c)\}\, \mathrm{d}x\, \mathrm{d}t - \int_{\mathbb{R}} \Phi(x, 0)\, V(u_0(x), c)\, \mathrm{d}x \leqq 0 \tag{18}$$

for all $c \in \mathbb{R}$ and for all nonnegative test functions $\Phi \in J$, then u is called the *entropy solution* of (15).

Hence, (18) is expected to be fulfilled by the entropy solution for all test elements $\hat{\Phi} := (\Phi, c)$ with

$$\hat{\Phi} \in \hat{J} := \{(\Phi, c) \mid \Phi \in J, \Phi \geqq 0, c \in \mathbb{R}\}\,. \tag{19}$$

As an abbreviation, (18) will be written as

$$B(\hat{\Phi})\, u \leqq 0 \quad \text{for all} \quad \hat{\Phi} \in \hat{J}\,. \tag{20}$$

(16) obviously determines the functions F uniquely if F is normalized by $F(0, c) = 0$:

$$F(u, c) = \int_0^u \frac{\mathrm{d}V(\alpha, c)}{\mathrm{d}\alpha} \frac{\mathrm{d}q(\alpha)}{\mathrm{d}\alpha}\, \mathrm{d}\alpha\,. \tag{21}$$

Examples are

$$V(u) = u^2/2 \quad \text{(independent of } c\text{)}. \tag{22}$$

Hence,

$$\hat{J} = \{\Phi \in J \mid \Phi \geqq 0\} \subset J .$$

An advantage of this choice is its applicability also to systems of conservation laws, at least if q is a gradient.

$$V(u, c) = |u - c| . \tag{23}$$

$$V(u, c) = \begin{cases} u - c, & u \geqq c, \\ 0, & u < c, \end{cases} \tag{24}$$

hence

$$F(u, c) = \begin{cases} q(u) - q(c), & u \geqq c, \\ 0, & u < c. \end{cases} \tag{25}$$

It should be mentioned that all these choices are equivalent in the following sense: Assume the weak solution u to jump from u_l to u_r if we cross a shock curve $\Gamma \subset \Omega$ from the left (l) to the right (r) parallel to the x-axis, and denote by $[\omega(u)] := \omega(u_l) - \omega(u_r) := \omega_l - \omega_r$ the jump of a function $\omega(u)$. Then (18) is equivalent with the jump condition,

$$[V(u, c)] \frac{[q(u)]}{[u]} \leqq [F(u, c)] \quad \text{for all} \quad c \in \mathbb{R} . \tag{26}$$

This follows from the fact that (15) is equivalent with the validity of (12) on subdomains where u is smooth together with the Rankine-Hugoniot condition

$$[u]\, s = [q(u)] \quad (s : \text{speed of the discontinuity})$$

(cf. e.g. [21], p. 248) and from the fact that similarly (18) is equivalent with

$$[V(u, c)] \cdot s \leqq [F(u, c)] \quad \text{for all} \quad c \in \mathbb{R} .$$

But for all our choices V, (26) is equivalent with

$$u_l \geqq u_r \tag{27}$$

because of the convexity of q. This can easily be checked in case of (i) and can also easily be verified in case of (ii) and (iii) if one takes the different cases

$$u_l, u_r \geqq c; \qquad u_l, u_r < c; \qquad u_l \geqq u_r < c$$

etc. carefully into account.

Concerning examples for the numerical treatment of (15), we restrict ourselves to explicit 3-point difference schemes of conservation form, namely

$$\frac{u_j^{\nu+1} - u_j^\nu}{\Delta t} + \frac{1}{\Delta x}\{g_{j+(1/2)}^\nu - g_{j-(1/2)}^\nu\} = 0 , \tag{28}$$

where Δt, Δx are step sizes in time- and space-direction, respectively, where u_j^ν is expected to be an approximation of $u(x_j, t_\nu)$ with $x_j = j \cdot \Delta x$, $j = 0, \pm 1, \pm 2, \ldots, t_\nu = \nu\, \Delta t$, $\nu = 0, 1, 2, \ldots$, and where $g_{j+(1/2)}^\nu$ abbreviates

$$g_{j+(1/2)}^\nu := g(u_j^\nu, u_{j+1}^\nu)$$

with a lipschitz continuous *numerical flux* $g : \mathbb{R} \times \mathbb{R} \to \mathbb{R}$. In order to make (28) an approximation of (12), g is assumed to fit the *consistency condition*

$$g(\sigma, \sigma) = q(\sigma) \quad \text{for all} \quad \sigma \in \mathbb{R} . \tag{29}$$

The discrete initial values are

$$u_j^0 := \frac{1}{\Delta x} \int\limits_{(j-(1/2))\Delta x}^{(j+(1/2))\Delta x} u_0(\xi)\, \mathrm{d}\xi . \tag{30}$$

Let the step ratio fulfill a Courant-Friedrichs-Lewy condition (C.F.L. condition):

$$0 < \lambda := \Delta t/\Delta x = \text{const} \leqq 1/|q'|_\infty^* \tag{31}$$

(cf. (14)). Assume $\Delta x = O(1/n)$, $n \in \mathbb{N}$, and put

$$u_n(x, t) = u_j^\nu \quad \text{for} \quad \begin{cases} (j - \frac{1}{2})\, \Delta x \leqq x < (j + \frac{1}{2})\, \Delta x, & j = 0, \pm 1, \pm 2, \ldots, \\ \nu\, \Delta t \leqq t < (\nu + 1)\, \Delta t, & \nu = 0, 1, 2, \ldots \end{cases} \tag{32}$$

If X_n denotes the space of functions $f\colon \mathbb{R}\times\mathbb{R}\to\mathbb{R}$ constant on the rectangles mentioned on the right hand side of (32), obviously $u_n\in X_n\subset X$, and (28), or equivalently

$$u_j^{\nu+1}-u_j^\nu+\lambda\{g_{j+(1/2)}^\nu-g_{j-(1/2)}^\nu\}=0\,, \tag{33}$$

can be written as

$$\hat{A}_n u_n=0\,,\qquad n=1,2,\dots, \tag{34}$$

with operators $\hat{A}_n\colon X_n\to\mathbb{R}$.

Because of the explicitness of the method, there is exactly one approximate solution u_n for every fixed $n\in\mathbb{N}$ such that

$$S_n:=\{u_n\in X_n\mid u_n \text{ solves (34)}\}\neq\emptyset\,.$$

It follows immediately from (28), that $u_n\in S_n$ fulfills also

$$A_n(\Phi)\,u_n:=\frac{1}{\Delta t}\iint\limits_\Omega\{u_n(x,t+\Delta t)-u_n(x,t)\}\,\Phi(x,t)\,\mathrm{d}x\,\mathrm{d}t$$

$$+\frac{1}{\Delta x}\iint\limits_\Omega\{g(u_n(x,t),u_n(x+\Delta x,t))-g(u_n(x-\Delta x,t),u_n(x,t))\}\,\Phi(x,t)\,\mathrm{d}x\,\mathrm{d}t=0\quad\text{for all}\quad\Phi\in J\,. \tag{35}$$

Hence, as in the elliptic example of section 2, every solution of the approximate problem $\hat{A}_n u_n=0$ is at the same time a solution of the weakly formulated approximate problem $A_n(\Phi)\,u_n=0$ for all $\Phi\in J$.

If we introduce

$$V_j^\nu(c):=V(u_j^\nu,c)$$

with the entropy function from (16), (17), we call a numerical solution a *discrete entropy solution* if

$$\frac{V_j^{\nu+1}(c)-V_j^\nu(c)}{\Delta t}+\frac{G_{j+(1/2)}^\nu(c)-G_{j-(1/2)}^\nu(c)}{\Delta x}\leqq 0\quad\text{for all}\quad c\in\mathbb{R}\,, \tag{36}$$

where $G_{j+(1/2)}^\nu(c):=G(u_j^\nu,u_{j+1}^\nu,c)$ with a *numerical entropy flux* G to be defined which fulfills the consistency condition

$$G(\sigma,\sigma,c)=F(\sigma,c)\quad\text{for all}\quad\sigma\in\mathbb{R}\quad\text{and all}\quad c\in\mathbb{R} \tag{37}$$

with F from (21).

The example

$$G(\alpha,\beta,c):=F_+(\alpha,c)+F_-(\beta,c) \tag{38}$$

with

$$F_+(\alpha,c):=\begin{cases}F(\alpha,c)\,, & \alpha\geqq 0\,,\\ 0\,, & \alpha<0\,,\end{cases}\qquad F_-(\beta,c):=\begin{cases}0\,, & \beta\geqq 0\,,\\ F(\beta,c)\,, & \beta<0\,,\end{cases}$$

obviously fulfills this consistency condition.

The weak formulation of (36) arises from multiplication of (36) by nonnegative functions $\Phi\in J$ and integration:

$$\iint\limits_\Omega\Phi(x,t)\left\{\frac{V(u_n(x,t+\Delta t),c)-V(u_n(x,t),c)}{\Delta t}\right.$$

$$\left.+\frac{G(u_n(x,t),u_n(x+\Delta x,t),c)-G(u_n(x-\Delta x,t),u_n(x,t),c)}{\Delta x}\right\}\mathrm{d}x\,\mathrm{d}t\leqq 0\quad\text{for all}\quad\hat{\Phi}\in\hat{J}\,.$$

Integration by parts and some evident coordinate shifts yield the weakly formulated discrete entropy condition

$$B_n(\hat{\Phi})\,u_n\leqq 0\quad\text{for all}\quad\hat{\Phi}\in\hat{J} \tag{39}$$

with

$$B_n(\hat{\Phi})\,u_n:=$$

$$-\int\limits_{\Delta t}^{\infty}\int\limits_{\mathbb{R}}\frac{\Phi(x,t)-\Phi(x,t-\Delta t)}{\Delta t}\,V(u_n(x,t),c)\,\mathrm{d}x\,\mathrm{d}t-\frac{1}{\Delta t}\int\limits_0^{\Delta t}\int\limits_{\mathbb{R}}\Phi(x,t)\,V(u_n(x,t),c)\,\mathrm{d}x\,\mathrm{d}t$$

$$-\iint\limits_\Omega\frac{\Phi(x+\Delta x/2,t)-\Phi(x-\Delta x/2,t)}{\Delta x}\,G(u_n(x-\Delta x/2,t),u_n(x+\Delta x/2,t),c)\,\mathrm{d}x\,\mathrm{d}t\leqq 0\quad\text{for all}\quad\hat{\Phi}\in\hat{J}\,, \tag{40}$$

and every function u_n that fulfills (36) fulfills (39), too.

4. Discrete functional analysis of weakly formulated problems

Assume J to be an index set and X to be a topological space. Let Y be a topological space, $\{a(\Phi), \Phi \in J\} \subset Y$ a given set and $\{A(\Phi), \Phi \in J\}$ a set of operators each of which maps $D := \bigcap_{\Phi \in J} D(A(\Phi)) \subset X$ into Y; $D(A(\Phi))$ is the Domain of $A(\Phi)$. Consider the problem:

Find $u \in X$ such that

$$A(\Phi)\,u = a(\Phi) \quad \text{for all} \quad \Phi \in J\,. \tag{41}$$

The elements of $S := \{u \in X \mid u \text{ solves } (41)\}$ are called *(weak) solutions.*

Let (41) be numerically approximated by a sequence of problems

$$\hat{A}_n u_n = \hat{a}_n\,, \qquad n = 1, 2, \ldots, \tag{42}$$

where the mapping $\hat{A}_n$, $n = 1, 2, \ldots$, maps a subspace $X_n \subset X$ into Z with a certain topological space Z. $\{\hat{a}_n\} \subset Z$ is a given d-compact sequence.

Herewith, a sequence of sets $\{S_n\}$ is called d-*compact (discretely compact)* if, for all $\tilde{\mathbb{N}} \leqq \mathbb{N}$, every subsequence $\{u_n\}_{n \in \tilde{\mathbb{N}}}$ such that $u_n \in S_n$, $n \in \tilde{\mathbb{N}}$, has a convergent subsequence. In this context, sequences $\{\hat{a}_n\}$ are expounded as sequences of one element sets.

Let

$$S_n := \{u_n \in X_n \mid u_n \text{ solves } (42)\}\,, \qquad n = 1, 2, \ldots,$$

and assume

$$S_n \neq \emptyset\,. \tag{43}$$

The solutions u_n of (42) are called *approximate solutions* of (41).

Let $\{A_n(\Phi) \mid \Phi \in J\}$, $n = 1, 2, \ldots$, be a sequence of sets of operators with $A_n(\Phi): X_n \to Y$, and let $\{a_n\}$ be a sequence of given mappings from J into Y with

$$\lim_{n \to \infty} a_n(\Phi) = a(\Phi) \quad \text{for every fixed} \quad \Phi \in J\,. \tag{44}$$

Assume $u_n \in S_n$ to fulfill also

$$A_n(\Phi)\,u_n = a_n(\Phi) \quad \text{for all} \quad \Phi \in J\,. \tag{45}$$

(45) is called the *weaks formulation of the approximate problem.*

We need three basic definitions.

Definition 1: A pair $[\{C_n\}, C]$ consisting of a sequence $\{C_n\}$ of operators and of an operator C is called *discretely closed* if the following implication holds:

$$u_n \to u \wedge C_n u_n \to z \Rightarrow Cu = z\,. \tag{46}$$

Definition 2: A sequence $\{C_n\}$ of operators is called *asymptotically regular* if the following implication holds:

$$\{C_n u_n\} \text{ d-compact} \Rightarrow \{u_n\} \text{ d-compact.} \tag{47}$$

Definition 3: Method (42) is called *convergent* if set convergence

$$S_n \to S$$

is ensured in the sense of the implication

$$\{u_n \mid u_n \in S_n, n \in \tilde{\mathbb{N}} \subset \mathbb{N}\} \to u \Rightarrow u \in S\,. \tag{48}$$

Concerning convergence of the numerical procedure, the following theorem holds.

Theorem 1: *Let*

(i) $[\{A_n(\Phi)\}, A(\Phi)]$ *be discretely closed for every fixed* $\Phi \in J$,
(ii) $\{\hat{A}_n\}$ *be asymptotically regular.*

Then

$$S_n \to S$$

in the sense of Definition 3 *and, hence,* $S \neq \emptyset$ *(existence of a solution of* (41)).

Proof: $\{\hat{a}_n\}$ d-compact $\Rightarrow \{\hat{A}_n u_0 \mid u_n \in S_n\}$ d-compact (in Z) $\Rightarrow \{u_n\}$ d-compact (in X) $\Rightarrow$ there exists $\{u_n \mid n \in \mathbb{N}' \subset \mathbb{N}\}$ and $u \in X$ with $u_n \to u (n \in \mathbb{N}')$. Herewith, $\mathbb{N}'$ and u are independent of Φ. Because of $A_n(\Phi)\, u_n = a_n(\Phi)$ and $a_n(\Phi) \to a(\Phi)$ for all $\Phi \in J$, (i) yields $A(\Phi)\, u = a(\Phi)$, i.e. $u \in S$.

This theorem generalizes slightly results presented in [3].

Remark 1: If one is only interested in particular right hand sides $a_n(\Phi)$ and $a(\Phi)$, (ii) can obviously be replaced by the weaker assumption

(ii)* $\{S_n\}$ d-compact.

Remark 2: If a uniqueness theorem holds for a particular problem of type (41), and if the assumptions of Theorem 1 are fulfilled, not only $S_n \to S$ but every full sequence $\{u_n \mid u_n \in S_n, n \in \mathbb{N}\}$ converges to the unique solution u.

If the solution (41) is not necessarily unique, problem (41) is sometimes completed by an additional inequality constraint of the type

$$B(\hat{\Phi})\, u \leqq 0 \quad \text{for all} \quad \hat{\Phi} \in \hat{J} \tag{49}$$

where $\{B(\hat{\Phi}) \mid (\hat{\Phi} \in \hat{J}\}$ is a set of nonlinear continuous functionals mapping X into $\mathbb{R}$ with a certain index set $\hat{J}$.

Because of applications in gas dynamics, (49) is called an *entropy condition* (cf. Section 3) and the unique solution $\hat{u}$ of $\{(41), (49)\}$ – if it exists – is called the *entropy solution*.

Assume a uniqueness theorem to be true:

(41) *together with* (49) *has at most one solution.* (50)

This particulary holds in the case of scalar conservations laws and the entropy condition mentioned in Section 3. If the original weakly formulated problem (41) is completed by an entropy condition (49), let us also complete the approximate problem (42) by inequality constraints

$$B_n(\hat{\Phi})\, u_n \leqq 0 \quad \text{for all} \quad \hat{\Phi} \in \hat{J}\,, \qquad n = 1, 2, \ldots, \tag{51}$$

with continuous functionals $B_n(\hat{\Phi})$ and assume that $\{(42), (51)\}$ has at least one solution $\hat{u}_n \in S_n$ for every fixed $n \in \mathbb{N}$.

We call (51) a sequence of *discretized entropy conditions* but we do not necessarily expect that the *discrete entropy solutions* $\hat{u}_n$ are unique for fixed $n \in \mathbb{N}$.

Principles for the construction of suitable functionals $B_n(\hat{\Phi})$ will use the following definition.

Definition 4: A sequence $\{C_n\}$ of operators is called *continuously convergent* to an operator C ($C_n \overset{c}{\to} C$) if the implication

$$u_n \to u \Rightarrow C_n u_n \to Cu \tag{52}$$

is valid ([7], [9]).

Then the following theorem holds.

Theorem 2: *Let* $\hat{S}_n := \{\hat{u}_n \in S_n \mid \hat{u}_n$ fulfills (51)$\}$ *and assume*

$$B_n(\hat{\Phi}) \overset{c}{\to} B(\hat{\Phi}) \quad \textit{for every fixed } \hat{\Phi} \in \hat{J}\,.$$

Moreover, let the assumptions of Theorem 1 *be fulfilled.*

Then the entropy solution $\hat{u}$ *of* $\{(41), (49)\}$ *exists and*

$$\hat{S}_n \to \{\hat{u}\}\,,$$

i.e. evergy sequence $\{\hat{u}_n\}$, $\hat{u}_n \in \hat{S}_n$, $n \in \mathbb{N}$, *converges to the unique entropy solution of the original weakly formulated problem.*

Proof: Because of $S_n \to S$ (cf. Theorem 1), there is a subsequence $\{\hat{u}_n \mid \hat{u}_n \in \hat{S}_n, n \in \mathbb{N}' \subset \mathbb{N}\}$ and a weak solution $u \in S$ with

$$\hat{u}_n \to u\,, \qquad n \in \mathbb{N}'\,.$$

The continuous convergence yields $B_n(\hat{\Phi})\, \hat{u}_n \to B(\hat{\Phi})\, u$. Because of (51), $B(\hat{\Phi})\, u \leqq 0$. Hence, u is the entropy solution $\hat{u}$. Because $\hat{u}$ is unique, not only subsequences $\{\hat{u}_n \mid n \in \mathbb{N}'\}$ converge to $\hat{u}$ but every full sequence $\{\hat{u}_n \mid n \in \mathbb{N}\}$ behaves so.

Remark 3: We remind the reader of the fact that for individually continuous operators K and K_n, $n \in \mathbb{N}$, the continuous convergence $K_n \overset{c}{\to} K$ on a metric space X^* is equivalent with the simultaneous presence of the two properties

(i) $\{K_n\}$ is a sequence of equicontinuous operators,
(ii) $K_n \to K$ pointwise on a set $\tilde{X} \underset{\text{dense}}{\subset} X^*$

(cf. [20], p. 78).

Often, the topological space situation arises from a metric space situation such that (j) and (jj) can be considered as the essential properties of the functionals $B_n(\hat{\Phi})$ to be constructed.

5. Application to the elliptic example of section 2

It follows from Section 2 that problem (1) with its weak formulation (2), its numerical procedure (10), and its weakly formulated numerical procedure (6) fits the terminology of the general problem treated in Section 4.

We need two lemmata.

Lemma 1 (Mountain Pass Lemma, cf. [1]):
Let $f \in C^1(X, \mathbb{R})$, X Banach space. Suppose:

(i) *f satisfies a Palais-Smale condition* (P.S. *condition*), *i.e. $\{f(u_i)\}$ bounded for a sequence $\{u_i\} \subset X$, $\lim_{i\to\infty} f'(u_i) = 0$, $\Rightarrow \{u_i\}$ d-compact*

(ii) *$f(\theta) = 0$ (θ zero element in X), $f(u)|_{u \in \partial B_\varrho(\theta)} \geqq \alpha > 0$ where $B_\varrho(\theta) := \{u \in X \mid \|u\| \leqq \varrho\}$, $\alpha = \text{const}$, there exists $e \in X$ with $\|e\| > \varrho$ and $f(e) \leqq 0$.*

Then

$$c := \inf_{\Gamma} \max_{t \in [0,1]} f(k(t)) \geqq \alpha$$

is a critical value of f, i.e. there are some $u \in X$ with $f(u) = c$, $f'(u) = 0$. Herewith, k is a path from θ to e and

$$\Gamma := \{k \in C([0,1], X);\, k(0) = \theta,\, k(1) = e\}\,.$$

The next lemma is obviously true.

Lemma 2: *If we put*

$$I(u) := \|u\|^2/2 - \int_\Omega G(x, u(x))\,dx\,, \qquad u \in H_0^1(\Omega) = X\,, \tag{53}$$

then $I \in C^1(X, \mathbb{R})$, i.e. the Fréchet derivative of I at every $u \in X$ exists and is continuous. Moreover,

$$I'(u)\,\Phi = \int_\Omega (\nabla u \cdot \nabla \Phi - g(x,u)\,\Phi)\,dx\,,$$

i.e.

$$I'(u)\,\Phi = A(\Phi)\,u \quad \textit{for all} \quad u \in X \quad \textit{and all} \quad \Phi \in X = J \tag{54}$$

(cf. [11]).

As a trivial consequence, the restriction of I to the subspace X_n fulfills $I \in C^1(X_n, \mathbb{R})$, and equations (2), (8) are equivalent to

$$I'(u)\,\Phi = 0 \quad \text{for all} \quad \Phi \in X\,, \tag{55}$$

$$I'(u_n)\,\Phi^{(n)} = 0 \quad \text{for all} \quad \Phi^{(n)} \in X_n\,, \tag{56}$$

respectively.

Now we are going to prove that the Mountain Pass Lemma is fulfilled for I on every subspace X_n:

Obviously, $I(\theta) = 0$. In order to find $e \in X_n$, $e \notin X_n \cap B_\varrho(\theta)$, $I(e) \leqq 0$, choose an element $\Phi_1 \in X_1 \subset X$ with $\Phi_1(x) > 0$ and with

$$\int_\Omega \Phi_1^2(x)\,dx = 1$$

and denote

$$\varphi(t) := I(t\Phi_1) = \tfrac{1}{2} t^2 \|\Phi_1\|^2 - \int_\Omega G(x, t\Phi_1(x))\,dx\,, \qquad t > 0\,.$$

Let $\delta > \|\Phi_1\|^2$ and fix $\beta > 0$ such that $g(x,s) \geqq \delta s$ for $s \geqq \beta$ (cf. H4)). Then

$$\begin{aligned}\int_\Omega G(x, t\Phi_1(x))\,dx &= \int_\Omega \int_0^{t\Phi_1(x)} g(x,s)\,ds\,dx = \int_\Omega \int_\beta^{t\Phi_1(x)} g(x,s)\,ds\,dx + \int_\Omega \int_0^\beta g(x,s)\,ds\,dx \\ &\geqq \int_\Omega \int_\beta^{t\Phi_1(x)} g(x,s)\,ds\,dx - \int_\Omega \int_0^\beta |g(x,s)|\,ds\,dx \\ &= \int_{t\Phi_1 \geqq \beta} \int_\beta^{t\Phi_1(x)} g(x,s)\,ds\,dx - \int_{t\Phi_1 < \beta} \int_{t\Phi_1}^\beta g(x,s)\,ds\,dx - M'(\beta)\end{aligned}$$

$$M'(\beta) := \int_\Omega \int_0^\beta |g(x,s)|\,ds\,dx = \text{const.}$$

Because of

$$\operatorname{mes}\left\{x \in \Omega, \Phi_1(x) < \frac{\beta}{t}\right\} \to 0 \quad \text{for} \quad t \to \infty ,$$

$$\left| \int_{t\Phi_1 < \beta} \int_{t\Phi_1}^{\beta} g(x, s) \, ds \, dx \right| < \varepsilon' \quad \text{for} \quad t \text{ sufficiently large.}$$

Hence,

$$\int_{\Omega} G(x, t\Phi_1(x)) \, dx \geqq \int_{t\Phi_1 \geqq \beta} \int_{\beta}^{t\Phi_1(x)} \delta s \, ds \, dx - \varepsilon' - M'(\beta)$$

$$= \frac{\delta}{2} \int_{t\Phi_1 \geqq \beta} \{t^2\Phi_1^2 - \beta^2\} \, dx - \varepsilon' - M'(\beta)$$

$$= \frac{\delta}{2} \left[\int_{\Omega} \{t^2\Phi_1^2 - \beta^2\} \, dx + \int_{t\Phi_1 < \beta} \{\beta^2 - t^2\Phi_1^2\} \, dx \right] - \varepsilon' - M'(\beta)$$

$$\geqq \frac{\delta}{2} (t^2 - \beta^2 |\Omega|) - \varepsilon' - M'(\beta) = \frac{\delta}{2} t^2 - M''(\beta, \delta)$$

for such sufficiently large t and with a certain constant $M''(\beta, \delta)$. Thus,

$$\varphi(t) \leqq \tfrac{1}{2} t^2 \|\Phi_1\|^2 - \frac{\delta}{2} t^2 + M''(\delta, \beta) = \tfrac{1}{2} (\|\Phi_1\|^2 - \delta) t^2 + M''(\beta, \delta)$$

which yields

$$\varphi(t) \to -\infty \quad \text{for} \quad t \to \infty .$$

This leads to the existence of $t_0 > 0$ such that

$$e := t_0\Phi_1 \in X_1 \subset X_n \subset X$$

and $\|e\| > \varrho$, $I(e) \leqq 0$.

From the assumptions H2) and H3) of Section 2, it can easily be shown that

$$I(u)|_{\partial B_\varrho(\theta)} \geqq \alpha_0 > 0 , \qquad \alpha_0 = \text{const} \tag{57}$$

(cf. [1]) for certain small $\varrho > 0$, and then evidently

$$I(u_n)|_{\partial B_\varrho^{(n)}(\theta)} \geqq \alpha_0^{(n)} \geqq \alpha_0 > 0 \tag{58}$$

for every fixed $n \in \mathbb{N}$, where $B_\varrho^{(n)}(\theta) := X_n \cap B_\varrho(\theta)$.

In order to prove that $I(u_n)$, $u_n \in X_n$, satisfies the P.S. condition, it is sufficient to show that the implication

$$\left.\begin{array}{l} \{u_n^{(m)}\} \subset X_n \\ \{I(u_n^{(m)})\} \text{ bounded} \\ I'(u_n^{(m)}) \to 0 \text{ if } m \to \infty \end{array}\right\} \Rightarrow \{u_n^{(m)}\} \quad \text{bounded}$$

holds for every fixed $n \in \mathbb{N}$.

Indeed, if $|I(u_n^{(m)})| \leqq c_n$, $m = 1, 2, \ldots$, then from assumption H5)

$$c_n \geqq I(u_n^{(m)}) \geqq \|u_n^{(m)}\|^2/2 - \int_{|u_n^{(m)}(x)| \geqq M} G(x, u_n^{(m)}(x)) \, dx - M_1$$

$$\geqq \|u_n^{(m)}\|^2/2 - \vartheta \int_{\Omega} u_n^{(m)}(x) \, g(x, u_n^{(m)}(x)) \, dx - M_2$$

$$\geqq (1/2 - \vartheta) \|u_n^{(m)}\|^2 + \vartheta \int_{\Omega} \{(\nabla u_n^{(m)})^2 - u_n^{(m)}(x) \, g(x, u_n^{(m)}(x))\} \, dx - M_2$$

$$= (1/2 - \vartheta) \|u_n^{(m)}\|^2 + \vartheta I'(u_n^{(m)}) \, u_n^{(m)} - M_2$$

with some constants c_n, M_1, M_2, $0 < \vartheta < 1/2$.

Because of $I'(u_n^{(m)}) \to 0$, we find $|I'(u_n^{(m)}) \, u_n^{(m)}| \leqq \|u_n^{(m)}\|$ for m sufficiently great, and hence,

$$c_n \geqq (1/2 - \vartheta) \|u_n^{(m)}\|^2 - \vartheta \|u_n^{(m)}\| - M_2 . \tag{59}$$

Thus, $\{\|u_n^{(m)}\|\}$ bounded, and the P.S. condition is fulfilled for $I(u_n)$ since X_n is finite dimensional.

By the Mountain Pass Lemma, there exist elements $u_n \in X_n$ satisfying (56) and therefore also (8) for every fixed $n \in \mathbb{N}$, i.e. $S_n \neq \emptyset$.

We are now going to show that $\bigcup_{n \in \mathbb{N}} S_n$ is bounded.

In fact, from (58) we have

$$I_n := I(u_n) = \inf_{k \in \Gamma_n} \max_{0 \leq t \leq 1} I(k(t)) \geq \alpha_0 > 0 \quad \text{for all} \quad n \in \mathbb{N}$$

with

$$\Gamma_n = \{k \in C([0, 1], X_n) \mid k(0) = \theta, k(1) = e\}.$$

Obviously, $\Gamma_n \subset X_n \subset X$. Because of $X_n \subset X_{n+1}$, we have $\alpha_0 \leq I_{n+1} \leq I_n, n \in \mathbb{N}$. Hence, $\{I_n\}$ bounded: $|I_n| \leq \hat{c}$ for all $n \in \mathbb{N}$, with a certain constant $\hat{c}$, and u_n is a critical point of I if restricted on the subspace X_n. Hence,

$$\hat{c} \geq I_n = \|u_n\|^2/2 - \int_\Omega G(x, u_n(x))\, dx .$$

By the same arguments as used to obtain (59), $\{u_n\}$ is bounded uniformly, i.e. $\bigcup_{n \in \mathbb{N}} S_n$ bounded.

In order to complete our analysis, we have to point out that $[\{A_n(\Phi)\}, A(\Phi)]$ is discretely closed and $\{S_n\}$ d-compact, i.e. $S_n \to S$ (cf. Theorem 1 and Remark 1).

Since $\overline{\bigcup X_n} = X = H_0^1(\Omega)$, we have

$$P_n u \to u \quad \text{for all} \quad u \in X ,$$

and

$$u_n \rightharpoonup \tilde{u}_0 \quad \text{yields} \quad P_n u_n \rightharpoonup \tilde{u}_0 \quad \text{for all} \quad \tilde{u}_0 \in X .$$

It can easily be developed from assumptions H1), H2) that

$$(Au_n, \Phi)_1 = (u_n + Ku_n, \Phi)_1 \to (\tilde{u}_0 + K\tilde{u}_0, \Phi)_1$$

for $u_n \in X$ and $u_n \rightharpoonup \tilde{u}_0 \in X$ (cf. [14], [17]). Hence, $P_n A(\Phi)\, u_n \to A(\Phi)\, u_0$, i.e. $P_n A(\Phi) \overset{c}{\to} A(\Phi)$ for all $\Phi \in X$. This result includes

$$[\{P_n A(\Phi)\}, A(\Phi)] \quad \text{discretely closed}$$

for every fixed $\Phi \in X = J$ (cf. [3]).

Because of the implications

$$(10) \Leftrightarrow (9) \Leftrightarrow (8) \Leftrightarrow (7),$$

(5) and (7) yield

$$u_n + P_n K u_n = 0 , \qquad u_n \in S_n, n \in \mathbb{N} . \tag{60}$$

Because $\{u_n\}$ bounded, there is some $y_0 \in X$ such that $P_n K u_n \to y_0$. Together with (60), this shows

$$\{S_n\} \text{ d-compact} .$$

6. Application to the hyperbolic example of section 3

It follows from Section 3 that problem (12) with its weak formulation (15), its numerical procedure (33), its weakly formulated numerical procedure (35), its entropy condition (20), and its discretized entropy condition (40) fits the terminology of the general problem treated in Section 4.

To show the discrete closure of $[\{A_n(\Phi)\}, A(\Phi)]$, i.e. property (i) of Theorem 1, we have to show the validity of the implication

$$\left.\begin{array}{ll} u_n \to u & \text{(with respect to the } L_1^{\text{loc}}\text{-topology)} \\ A_n(\Phi)\, u_n \to y & (\text{in } \mathbb{R}) \end{array}\right\} \Rightarrow A(\Phi)\, u = y .$$

But in the context of our example, this is nothing other than the well known Lax-Wendroff-Theorem [16] the assumptions of which are completely satisfied by the hyperbolic problem of Section 3 and its 3-point approximation.

Just for convenience with respect to the concretization of the discretized entropy condition, we restrict ourselves to the ENGQUIST-OSHER scheme [10]. (Similar results can be obtained for TVD schemes as presented by HARTEN [12].) Also in this particular case, property (ii)* is fulfilled. Namely, the Engquist-Osher scheme

$$u_j^{\nu+1} = u_j^\nu - \lambda\{[q_-(u_{j+1}^\nu) + q_+(u_j^\nu)] - [q_-(u_j^\nu) + q_+(u_{j-1}^\nu)]\} \tag{61}$$

if applied to (12), i.e. to (15), in case of a strictly convex function q with minimum at $u = 0$ and with $q(0) = 0$, where

$$q_-(u) = \begin{cases} 0, & u \geqq 0, \\ q(u), & u < 0, \end{cases} \qquad q_+(u) = \begin{cases} q(u), & u \geqq 0, \\ 0, & u < 0, \end{cases} \tag{62}$$

is of the form (28) with

$$g_{j+(1/2)} = q_-(u_{j+1}) + q_+(u_j). \tag{63}$$

Because of the C.F.L. condition (31), $|u_j^\nu| \leqq \|u_0\|_{L_\infty}$ for all $\nu \geqq 0$, and the scheme is monotone, i.e. $u_j^{\nu+1}$ is a monotone nondecreasing function of the variables $u_{j-1}^\nu, u_j^\nu, u_{j+1}^\nu$. But monotone schemes form an L_1 contraction [13], and the solutions to monotone schemes extended to the intergrid points by (32), hence, converge to a limit in the sense of L_1^{loc} topology [8]. Thus, $\{S_n\}$ d-compact and Theorem 1 applies.

For convenience, let us restrict to the case $u_0 \geqq 0$. Then also $c > 0$ is sufficient. With the entropy function V from (24), hence F from (25), and with the numerical entropy flux G from (38), the discretized entropy condition (36) reads for $c > 0$ as

$$V(u_j^{\nu+1}, c) \leqq V(u_j^\nu, c) - \lambda\{F(u_j^\nu, c) - F(u_{j-1}^\nu, c)\} \tag{64}$$

because of the relation easily to be verified in this particular example,

$$G(u_j, u_{j+1}, c) = F(u_j, c). \tag{65}$$

As already mentioned in (38), the consistency condition (37) is fulfilled, and $G(\alpha, \beta, c)$ is lipschitz continuous with respect to the variables α, β ($\leqq \|u_0\|_{L_\infty}$):

$$|G(\tilde{\alpha}, \tilde{\beta}, c) - G(\alpha, \beta, c)| = |F(\tilde{\alpha}, c) - F(\alpha, c)| \leqq |q'|_\infty^* \, |\tilde{\alpha} - \alpha|. \tag{66}$$

(64) can be written as

$$V(u_j^{\nu+1}, c) \leqq \begin{cases} u_j^\nu - c - \lambda\{q(u_j^\nu) - q(u_{j-1}^\nu)\}, & c \leqq u_j^\nu, u_{j-1}^\nu, \\ u_j^\nu - c - \lambda\{q(u_j^\nu) - q(c)\}, & u_{j-1}^\nu < c \leqq u_j^\nu, \\ 0, & u_{j-1}^\nu, u_j^\nu < c, \\ \lambda\{q(u_{j-1}^\nu) - q(c)\}, & u_j^\nu < c \leqq u_{j-1}^\nu. \end{cases} \tag{67}$$

Taking the cases $u_j^{\nu+1} \geqq c$ and $u_j^{\nu+1} < c$ into account and combining them with all the compatible sign distributions

$$u_{j-1}^\nu < 0, \qquad u_j^\nu < 0, \qquad u_{j+1}^\nu \geqq 0,$$

$$u_{j-1}^\nu < 0, \qquad u_j^\nu \geqq 0, \qquad u_{j+1}^\nu \geqq 0. \quad \text{etc.}$$

one can easily show that (67) holds provided the C.F.L. condition (31) is satisfied.

In order to demonstrate this, we only treat two particular cases:

α) $u_j^{\nu+1} < c, \quad u_{j-1}^\nu \geqq 0, \quad u_j^\nu < 0, \quad u_{j+1}^\nu \geqq 0.$

Then the discrete entropy condition reads as

$$0 \leqq \begin{cases} 0, & 0 \leqq u_{j-1}^\nu < c, \\ \lambda[q_{j-1}^\nu - q(c)], & c \leqq u_{j-1}^\nu, \end{cases}$$

and this condition is certainly satisfied because of $q_{j-1} \geqq q(c)$ from the convexity of q.

β) $u_j^{\nu+1} \geqq c, \quad u_{j-1}^\nu < 0, \quad u_j^\nu < 0, \quad u_{j+1}^\nu \geqq 0.$

$u_j^{\nu+1} \geqq c$ yields $u_j^\nu + \lambda q_j^\nu \geqq c > 0$, hence

$$\lambda > -\frac{u_j^\nu}{q_j^\nu} = -\frac{0 - u_j^\nu}{q(0) - q(u_j^\nu)} = -\frac{1}{q'(\tilde{u}_j)} \quad \text{with} \quad u_j^\nu < \tilde{u}_j < 0$$

such that $\lambda > 1/|q'|_\infty^*$. But because of the C.F.L. condition, this case can't really occur.

The same results can be obtained for $c \leqq 0$. Because (36) is fulfilled for all $c \in \mathbb{R}$, also the weakly formulated discrete entropy condition (40) is satisfied for all $\hat{\Phi} \in \hat{J}$ globally, i.e. for all $\Delta t > 0$, with the functions V, G mentioned above.

With V from (22) and G from (38), STRACK [22] showed that the discrete entropy condition is locally fulfilled by the Engquist-Osher solution, i.e. for controlled step sizes Δt_n that have to be sufficiently small.

It is well known that an only slight modification of the proof of the Lax-Wendroff theorem leads to the fact that a consistent multipoint finite difference scheme the solution of which converges for $\Delta t \to 0$ to a L_1^{loc} limit and which fulfills the discrete entropy condition, generates approximations to the entropy solution of the original problem. Hence, this also holds for the Engquist-Osher scheme as already shown by ENGQUIST and OSHER in their original paper.

Just in order to demonstrate our general theory of Section 4, we are going to show $B_n(\hat{\Phi}) \xrightarrow{c} B(\hat{\Phi})$ for all $\hat{\Phi} \in \hat{J}$, and we will do that for consistent three point schemes with lipschitz continuous function G. The generalization to multi point schemes is obvious, and the Engquist-Osher scheme is a particular example.

Concerning continuous convergence, remember Remark 3. Because we have to show $B_n(\hat{\Phi}) \xrightarrow{c} B_n(\hat{\Phi})$ for every fixed $\hat{\Phi} \in \hat{J}$ with $\hat{\Phi} = (\Phi, c)$, $c \in \mathbb{R}$, $\Phi \in J$, $\Phi \geqq 0$, we choose X^* as the space of functions $u \in X$ restricted to supp Φ equipped with the L_1-metric on X^*. Hence, the topoloy on X^* introduced by this metric corresponds with the L_1^{loc}-topology on X. Thus, we can show $B_n(\hat{\Phi}) \xrightarrow{c} B(\hat{\Phi})$ by realization of Remark 3 with its conditions (j) and (jj).

Proof of (j): For fixed $\hat{\Phi} \in J$, (40) yields

$$|B_n(\hat{\Phi})\,\tilde{u} - B_n(\hat{\Phi})\,u| = \Bigg|\int_{\Delta t}^{\infty}\int_{\mathbb{R}} \Phi_t(x, t - \vartheta\,\Delta t)\,[\,V(\tilde{u}(x,t),c) - V(u(x,t),c)]\,\mathrm{d}x\,\mathrm{d}t$$

$$+ \frac{1}{\Delta t}\int_0^{\Delta t}\int_{\mathbb{R}} \Phi(x,t)\,[V(\tilde{u}(x,t),c) - V(u(x,t),c)]\,\mathrm{d}x\,\mathrm{d}t$$

$$+ \iint_{\Omega} \Phi_x\left(x + \hat{\vartheta}\frac{\Delta x}{2}, t\right)[G(\tilde{u}(x - \Delta x/2, t), \tilde{u}(x + \Delta x/2, t), c)$$

$$- G(u(x - \Delta x/2, t), u(x + \Delta x/2, t), c)\,\mathrm{d}x\,\mathrm{d}t\Bigg|$$

with $0 < \vartheta = \vartheta(t) < 1$, $-1 < \hat{\vartheta} = \hat{\vartheta}(t) < 1$.

The compact support $\Omega^* \subset \Omega$ of Φ is also the support of the continuous functions Φ_x, Φ_t. Let K be a joint bound of Φ, Φ_x, Φ_t. Hence,

$$|B_n(\hat{\Phi})\,\tilde{u} - B_n(\hat{\Phi})\,u| \leqq K \iint_{\Omega^*} |V(\tilde{u}(x,t),c) - V(u(x,t),c)|\,\mathrm{d}x\,\mathrm{d}t$$

$$+ \frac{K}{\Delta t}\int_0^{\Delta t}\int_{\mathbb{R}} |V(\tilde{u}(x,t),c) - V(u(x,t),c)|\,\mathrm{d}x\,\mathrm{d}t$$

$$+ KL \iint_{\Omega^*} \{|\tilde{u}(x - \Delta x, t) - u(x - \Delta x, t)|$$

$$+ |\tilde{u}(x + \Delta x, t) - u(x + \Delta x, t)|\}\,\mathrm{d}x\,\mathrm{d}t\,.$$

Because of

$$|V(\tilde{u}, c) - V(u, c)| \leqq |\tilde{u} - u|$$

(cf. (24)), we find

$$|B_n(\hat{\Phi})\,\tilde{u} - B_n(\Phi)\,u| \leqq 2K(1 + L)\,\|\tilde{u} - u\|_{L_1(\Omega^*)}$$

with the Lipschitz constant L of G, i.e. equicontinuity.

Proof of (jj): For $u \in X^*$, (20) together with the consistency property of G and with (40) yields for every fixed $\hat{\Phi} \in \hat{J}$:

$$|B(\hat{\Phi})\,u - B_n(\hat{\Phi})\,u| = \Bigg|\iint_{\Omega} \{\Phi_t(x,t)\,V(u(x,t),c) + \Phi_x(x,t)\,F(u(x,t),c)\}\,\mathrm{d}x\,\mathrm{d}t$$

$$+ \int_{\mathbb{R}} \Phi(x,0)\,V(u_0(x),c)\,\mathrm{d}x - \int_{\Delta t}^{\infty}\int_{\mathbb{R}} \Phi_t(x, t - \vartheta\,\Delta t)\,V(u(x,t),c)\,\mathrm{d}x\,\mathrm{d}t$$

$$- \frac{1}{\Delta t}\int_0^{\Delta t}\int_{\mathbb{R}} \Phi(x,t)\,V(u(x,t),c)\,\mathrm{d}x\,\mathrm{d}t$$

$$- \iint_{\Omega} \Phi_x(x - \hat{\vartheta}\,\Delta x, t)\,G\left(u\left(x - \frac{\Delta x}{2}, t\right), u\left(x + \frac{\Delta x}{2}, t\right), c\right)\mathrm{d}x\,\mathrm{d}t\Bigg|$$

$$
\begin{aligned}
&= \Bigg| \int_0^{\Delta t} \int_{\mathbb{R}} \Phi_t(x,t)\, V(u(x,t),c)\, dx\, dt \\
&\quad + \int_{\Delta t}^{\infty} \int_{\mathbb{R}} \{\Phi_t(x,t) - \Phi_t(x, t - \vartheta\, \Delta t)\}\, V(u(x,t),c)\, dx\, dt \\
&\quad + \iint_{\Omega} \Big\{ \Phi_x(x,t)\, G(u(x,t), u(x,t), c) \\
&\qquad - \Phi_x(x - \hat{\vartheta}\, \Delta x, t)\, G\left(u\left(x - \frac{\Delta x}{2}, t\right), u\left(x + \frac{\Delta x}{2}, t\right), c\right)\Big\}\, dx\, dt \\
&\quad - \int_{+0}^{\Delta t} \int_{\mathbb{R}} \frac{\Phi(x,t)\, V(u(x,t),c) - \Phi(x,0)\, V(u(x,0),c)}{t} \frac{t}{\Delta t}\, dx\, dt \Bigg| . \qquad (68)
\end{aligned}
$$

$\tilde{X} := X^* \cap C_0^1(\Omega)$ is dense in X^* with respect to the L_1-Norm. Obviously, for fixed $\hat{\Phi} \in \hat{J}$, we get for $u \in \tilde{X}$,

$$\left| \int_0^{\Delta t} \int_{\mathbb{R}} \Phi_t(x,t)\, V(u(x,t),c)\, dx\, dt \right| \leqq \Delta t \cdot K(K^* + c)\, |\Omega^*| =: \varkappa\, \Delta t \qquad (69)$$

with the same constant K as before, with $K^* = \max_{\Omega^*} |u(x,t)|$, and with $|\Omega^*| = \iint_{\Omega^*} dx\, dt$.

Because Φ_t is uniformly continuous on Ω^*, we find for every $\varepsilon > 0$ a $\delta(\varepsilon) > 0$ such that

$$\left| \int_{\Delta t}^{\infty} \int_{\mathbb{R}} \{\Phi_t(x,t) - \Phi_t(x, t - \vartheta\, \Delta t)\}\, V(u(x,t),c)\, dx\, dt \right| \leqq (K^* + c)\, |\Omega^*|\, \varepsilon \quad \text{for all} \quad \Delta t < \delta(\varepsilon) . \qquad (70)$$

Let $\Delta t < \min(\varepsilon, \delta(\varepsilon))$. Moreover, taking the consistency property of G and its lipschitz continuity as well as the boundedness of Φ_x and the uniform continuity of Φ_x and of $u \in \tilde{X}$ on Ω^* into account, we get

$$
\begin{aligned}
&|\Phi_x(x,t)\, G(u(x,t), u(x,t), c) - \Phi(x - \hat{\vartheta}\, \Delta x, t)\, G(u(x - \Delta x/2, t), u(x + \Delta x/2, t), c)| \\
&\leqq |\Phi_x(x,t) - \Phi_x(x - \hat{\vartheta}\, \Delta x, t|\, |F(u(x,t),c)| \\
&+ |\Phi_x(x - \hat{\vartheta}\, \Delta x, t)| \cdot L\{|u(x,t) - u(x - \Delta x/2, t)| + |u(x,t) - u(x + \Delta x/2, t)|\} \\
&\leqq \sup_{\Omega^*} |F(u(x,t),c)|\, \varepsilon + 2LK\varepsilon \quad \text{for} \quad \Delta x, \Delta t \quad \text{sufficiently small.} \qquad (71)
\end{aligned}
$$

(68), (69), (70), (71) lead to

$$
\begin{aligned}
|B(\hat{\Phi})\, u - B_n(\hat{\Phi})\, u| &\leqq \varkappa \cdot \varepsilon + (K^* + c)\, |\Omega^*|\, \varepsilon \\
&+ \left\{ \sup_{\Omega^*} |F(u(x,t),c)|\, \varepsilon + 2LK\varepsilon \right\} |\Omega^*| + \int_0^{\Delta t} \left| \int_{\mathbb{R}} [\Phi(\cdot,\cdot)\, V(u(\cdot,\cdot),c)]_t\, (x, \tilde{\vartheta} t)\, dx \right| dt ,
\end{aligned}
$$

$0 < \tilde{\vartheta} < 1$, for all sufficiently small Δx, Δt.

But $[\Phi V(u,c)]_t$ is piecewise continuous for $u \in \tilde{X}$ and, thus, bounded on Ω^* by a certain constant $\tilde{K}$. Hence, with a certain constant Q,

$$|B(\hat{\Phi})\, u - B_n(\hat{\Phi})\, u| < Q\varepsilon$$

for all sufficiently small Δx, Δt, i.e. for all sufficiently great $n \in \mathbb{N}$, i.e. $B_n(\hat{\Phi})\, u \to B(\hat{\Phi})\, u$ for every fixed $\hat{\Phi} \in \hat{J}$ and for every $u \in \hat{X}$. This ends the proof of $B_n(\hat{\Phi}) \xrightarrow{c} B(\hat{\Phi})$.

Acknowledgement: The Research of Professor LEI was supported by Volkswagen Foundation and Science Foundation of China.

Literatur

1 AMBROSETTI, A.; RABINOWITZ, P. H.: Dual variational method in critical point theory and applications. J. Funct. Anal. **14** (1973), 349–381.

2 ANSELONE, P. M.; ANSORGE, R.: A unified framework for the discretization of nonlinear operator equations. Numer. Funct. Anal. Optimiz. **4** (1981), 61–99.

3 ANSELONE, P. M.; ANSORGE, R.: Discrete closure and asymptotic (quasi-) regularity in discretization algorithms. IMA J. Numer. Anal. **7** (1987), 431–448.

4 ANSELONE, P. M.; LEI, JINGAN: The approximate solution of monotone nonlinear operator equations. Rocky Mt. J. Math. **16** (1986), 791–801.
5 ANSORGE, R.: General principles for the convergence of discretization methods if applied to weakly formulated problems. In STREHMEL, K., (ed.): Proc. NUMDIFF-5 Teubner Leipzig (to appear).
6 ANSORGE, R.: What does the entropy condition mean in traffic flow theory? Transpn. Res.-B **24B** (1990),133–143.
7 COURANT, R.: Über eine Eigenschaft der Abbildungsfunktion bei konformer Abbildung. Göttinger Nachr. 1914, pp. 101–109.
8 CRANDALL, M. G.; MAJDA, A.: Monotone difference approximations for scalar conservation laws. Math. Comput. **34** (1980), 1–21.
9 DU BOIS-RAYMONT, P.: Sitzungs-Ber. Preuß. Akad. d. Wiss. Berlin 1886, pp. 359–360.
10 ENGQUIST, B.; OSHER, S.: Stable and entropy satisfying approximations for transonic flow calculations. Math. Comput. **34** (1980), 321–351.
11 GRIGORIEFF, R. D.: Über diskrete Approximationen nichtlinearer Gleichungen 1. Art. Math. Nachr. **69** (1975), 253–272.
12 HARTEN, A.: On a class of high resolution total-variation-stable finite-difference schemes. SIAM J. Numer. Anal. **21** (1984), 1–23.
13 HARTEN, A.; HYMANN, J. M.; LAX, P. D.: On finite difference approximations and entropy conditions for shocks. Comm. Pure Appl. Math. **29** (1976), 297–322.
14 KRASNOSELSKII, M. A.: Topological methods in the nonlinear integral equations. Pergamon Press, Oxford 1964.
15 LAX, P. D.: Shock waves and entropy. In ZARANTELLO, E. M. (ed): Contributions to nonlinear functional analysis. Academic Press, New York 1971, pp. 603–634.
16 LAX, P. D.; WENDROFF, B.: Systems of conservation laws, Comm. Pure Appl. Math. **13** (1960), 217–237.
17 LEI, JINGAN; ZENG, ZHONGGANG: The convergence of Galerkin solutions to nonlinear variational problems. J. Wuhan Univ. **4** (1986), 8–16 (in Chinese).
18 OLEINIK, O.: On the uniqueness of the generalized solution of the Cauchy problem for a nonlinear system of equations occurring in mechanics. Usp. Mat. Nauk (N.S.), **12** (1957), 169–176 (in Russian).
19 REINHARDT, H. J.: Analysis of approximation methods for differential and integral equations. Springer, New York 1985.
20 RINOW, W.: Die innere Geometrie der metrischen Räume. Springer, Berlin 1961.
21 SMOLLER, S.: Shock waves and reaction-diffusion equations. Springer, New York 1983.
22 STRACK, K. G.: Discrete entropy-condition and stability for conservation laws. Report No 30, Inst. für Geom. und Prakt. Mathematik der RWTH Aachen 1985.
23 STUMMEL, F.: Discrete convergence of mappings. In MILLER, J. J. (ed.): Proc. Conf. Num. Anal. Dublin, Academie Press 1973, pp. 285–310.
24 VAINIKKO, H. J.: Funktionalanalysis der Diskretisierungsmethoden. Teubner, Leipzig 1976.

Received September 17, 1990

Addresses: Prof. Dr. RAINER ANSORGE, Institut für Angewandte Mathematik der Universität Hamburg, Bundesstraße 55, W-2000 Hamburg 13, Germany; Prof. Dr. LEI JINGAN, Department of Mathematics, Wuhan University, Wuhan 430 072, P. R. China

BOOK REVIEWS

Scharlau, W. (Hgb.), Mathematische Institute in Deutschland 1800–1945. Braunschweig/Wiesbaden, Friedr. Vieweg & Sohn 1990. VI, 291 S., DM 72,–. ISBN 3-528-08992-X (Dokumente zur Geschichte der Mathematik 5)

Dem Herausgeber gebührt Dank, mit dem vorliegenden Band das erste Nachschlagewerk über die mathematischen Institute in Deutschland für den Zeitraum 1800–1945 erstellt zu haben, das für die mathematikhistorische Forschung bisher fehlte. Dabei standen dem Herausgeber eine Vielzahl von Mitarbeitern für einzelne Universitäten und Hochschulen zur Verfügung.

Jeder Abschnitt ist in vier Teile untergliedert. Der erste bringt eine kurze Geschichte der jeweiligen Universität, der zweite beschreibt die Entwicklung der mathematischen Forschung und Lehre dort. Im dritten Abschnitt werden die fest angestellten planmäßigen Professoren aufgelistet, ergänzt durch Lebensdaten und die Zeiten ihrer Wirksamkeit am betreffenden Ort. Hier mußte der Herausgeber mehr oder weniger ad hoc vorgehen, denn die Personalstruktur an den deutschen Universitäten war kompliziert, und in den einzelnen deutschen Staaten sowie in verschiedenen Zeitepochen waren die Titel für die Lehrbeauftragten recht unterschiedlich. Manche Jahreszahl in diesem Abschnitt ist ergänzungsbedürftig; teilweise sind unsichere Angaben schon mit Fragezeichen versehen; Druckfehler dürften bei einem so umfangreichen Werk nicht auszuschließen sein (z. B. S. 192 bei HOHEISEL, S. 67 bei ERHARD SCHMIDT). Im vierten Abschnitt „Habilitationen, Privatdozenten, Lehrbeauftragte“ wird die Liste vom Abschnitt 3 um die nicht planmäßig angestellten Wissenschaftler ergänzt.

Im Vorwort begründet der Herausgeber die Auswahl der behandelten Universitäten und Hochschulen. Insbesondere beschränkt er sich auf Städte innerhalb des Deutschen Bundes bzw. Deutschen Reiches, und es finden leider die Universitäten keine Beachtung, die vor oder während der napoleonischen Zeit geschlossen wurden (z. B. Erfurt, Helmstedt, Altdorf). Bei Breslau wird Frankfurt/O nicht behandelt, während bei Halle die Entwicklung in Wittenberg bis 1817 beschrieben wird.

Den Abschluß des Bandes, der für die mathematikhistorische Forschung von großer Bedeutung werden dürfte, bildet ein interessanter Aufsatz von G. SCHUBRING über die strukturelle Entwicklung der Mathematik an den deutschen Hochschulen für den behandelten Zeitraum.

Berlin D. ULLMANN

Nakagawa, T.; Ishikawa, H.; Tsurui, A. (eds.), Recent Studies on Structural Safety and Reliability. London/New York, Elsevier Applied Science 1989. XIII, 230 pp., 116 figs., £ 42.00. ISBN 1-85166-281-2 (Current Japanese Materials Research 5)

Der vorliegende 5. Band einer von der Gesellschaft für Materialwissenschaften Japans herausgegebenen Reihe des Elsevier-Verlages über Fortschritte auf den Gebieten Technologie und Materialwirtschaft in Japan behandelt Fragen von Zuverlässigkeit und Sicherheit mechanischer Systeme. Im Rahmen von 14 Fachbeiträgen werden von insgesamt 27 Autoren ausgewählte Probleme der Zuverlässigkeit unter besonderer Berücksichtigung stochastischer Einflüsse diskutiert. Einige wichtige Schwerpunktthemen sind stochastisches Rißwachstum, statistische Untersuchungen bei Materialermüdung, stochastische Analyse der Festigkeit ausgewählter Verbundwerkstoffe, bruchmechanische Aspekte spezieller keramischer Werkstoffe, Analyse von Versagensmechanismen in Verbindung mit Identifikationsverfahren, verschiedene Probleme der Systemzuverlässigkeitsbestimmung auf der Grundlage unscharfer Mengen, Zuverlässigkeitsanalyse durch Verkehrslasten erregter Brückenkonstruktionen basierend auf Niveauüberschreitungswahrscheinlichkeiten stochastischer Prozesse, Rißwachstum und Erneuerungstheorie, Instandhaltungsprobleme, ausgewählte Softwareentwicklungen für Zuverlässigkeitsprobleme sowie spezielle Fragen der Erdbebenzuverlässigkeit. Der an der Nutzung stochastischer Methoden in der Mechanik interessierte Leser findet zahlreiche sehr gute Arbeiten mit wichtigen neuen Ideen und Anregungen für Zuverlässigkeitsuntersuchungen verschiedenster Art.

Chemnitz C. LANGE

Roberts, J. B.; Spanos, P. D., Random Vibration and Statistical Linearization. Chichester etc., John Wiley & Sons 1990. XIII, 407 pp., £ 37.50. ISBN 0-471-91699-4

Für Zuverlässigkeitsuntersuchungen im Maschinenbau und Bauwesen ist die Theorie stochastischer Schwingungen von großer Wichtigkeit. Eine praktisch notwendige wirklichkeitsnahe Beschreibung der Vorgänge (z. B. infolge Erdbeben oder Wind stochastisch schwingende Bauwerke u. v. a. m.) erfordert die Einbeziehung nichtlinearer Effekte in die Modellierung. Mathematisch sind dann nichtlineare stochastische Differentialgleichungen verschiedensten Typs zu untersuchen. Eine gut geeignete Näherungsmethode ist hierzu die statistische Linearisierung. Nach einer anschaulichen Einführung werden Fragen der Modellbildung in Abhängigkeit vom zugrunde liegenden mechanischen System diskutiert. Erforderliche Grundlagen aus der Wahrscheinlichkeitstheorie werden bereitgestellt. Der Behandlung linearer Schwingungssysteme folgen Ausführungen zur statistischen Linearisierung ein- und mehrdimensionaler nichtlinearer Systeme mit stationärem Ausgangsprozeß. Ausgewählte nichtstationäre Probleme, Systeme mit (nichtlinearem) Hysteresecharakter, der Fall nicht notwendig gaußscher Erregung, die Berechnung von Versagenswahrscheinlichkeiten sowie Genauigkeitsuntersuchungen runden die behandelte Thematik ab. Die Ausführungen der Autoren sind für Ingenieure mit Vorkenntnissen auf dem Gebiet der Stochastik gut verständlich geschrieben und durch zahlreiche anschauliche Praxisbeispiele illustriert.

Chemnitz C. LANGE

Bluman, G. W.; Kumei, S., Symmetries and Differential Equations. Berlin etc., Springer-Verlag 1989. XIII, 412 pp., 21 figs., DM 114,00. ISBN 3-540-96996-9 (Applied Mathematical Sciences 81)

Diese Monographie will die Grundlagen der Anwendung der Lie-Gruppen bei Differentialgleichungen vermitteln. Zuerst wird eine Einführung in die Transformationsgruppen gegeben. Dabei verzichten die Autoren auf topologische Begriffe wie die der Mannigfaltigkeit und des Tangentialraumes und somit auch auf die geometrische Interpretation der Lie-Gruppe. Die Autoren legen den Begriff der einparametrigen Lie-Gruppe diesem Kapitel zugrunde. Darauf aufbauend wird die Invarianz einer gewöhnlichen Differentialgleichung bzgl. einer einparametrigen Lie-Gruppe definiert. Sie ist gegeben, falls diese Gruppe die Lösungsfamilie in sich abbildet. Dieses sind die grundlegenden Begriffe. Die Beispiele handeln im wesentlichen auch von einparametrigen Gruppen: R oder SO(2). Die Verfasser demonstrieren, wie man Invariante einer gewöhnlichen oder partiellen Differentialgleichung findet und dann mittels Integration die Ordnung der ursprünglichen Gleichungen reduziert.

Der Schwerpunkt des Buches liegt m. E. in seinen konstruktiven Methoden zur Berechnung von Lösungen vor allem von Differentialgleichungen der Mathematischen Physik. Auch die Modellierung physikalischer Prozesse und deren Dimensionsanalyse nimmt einen wichtigen Platz ein.

Berlin M. KERN

ZAMM · Z. angew. Math. Mech. **71** (1991) 7/8, 223–231 Akademie Verlag

BOHL, E.

A Boundary Layer Phenomenon for Linear Systems with a Rank Deficient Matrix

Dedicated to the memory of LOTHAR COLLATZ

Diese Arbeit betrachtet lineare Gleichungssysteme mit einer singulären Matrix und leitet eine Regularisierung her, um alle Lösungen zu berechnen. Im Falle, daß das System eine Störung einer Matrix mit weiterem Rangabfall ist, kann sich möglicherweise eine Grenzschicht ausbilden. Solche Systeme treten in der Enzymkinetik auf. Ein Beispiel dieser Art und numerische Resultate werden angegeben.

This paper considers linear systems with a singular matrix and provides a regularization to compute all solutions. If the system is a perturbation of a rank deficient matrix, a boundary layer phenomenon is possible. Systems of the kind considered occur in enzyme kinetics. A particular example and numerical results are given.

В статье рассматриваются линейные системы сингулярной матрицы и даются регуляризация для вычисления всех решений. Если система является возмущением ранговой неполной матрицы, потом явление пограничного слоя может встретиться. Системы рассмотренного рода возникают в кинетике энзима. Даются особый пример и численные результаты.

MSC (1980): 65F05, 65F18

0. Introduction

0.1. In a chemical network a unit of the form given in the upper part of figure 1 to produce amplification of an outside stimulus U to the chemical X_1 is quite common [3, 4, 6]. The unit is part of a complicated network which, among other things, creates a direct path between the chemicals X_1 and X_3. Note that there is no connection between X_1 and X_3 if $U = 0$. Therefore, there is a fundamental difference between $U = 0$ (no outside stimulus) and $U > 0$ (existing stimulus,

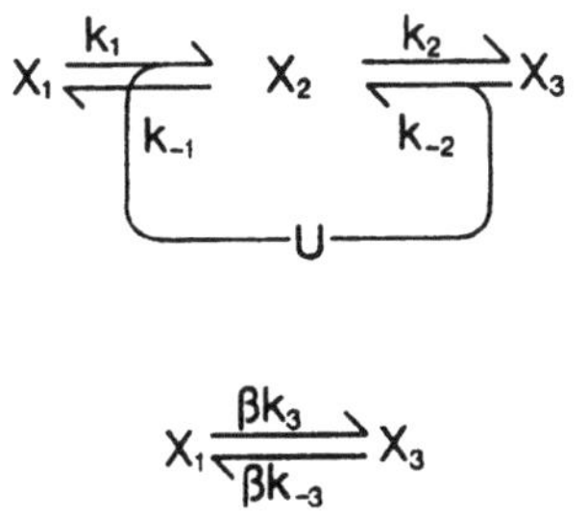

Fig. 1. A model-network

possibly very small). The embedding into a more complicated network typically defines a nonlinear action creating the bridge between X_1 and X_3. We try to simulate this bridge in the most simple (linear) way adding the first order reaction given in the lower part of figure 1. Here, $\beta \geqq 0$ is a control parameter and $k_i, k_{-i} > 0$ are the respective rate constants. We denote by $x_j(t)$, u the concentration of the chemicals X_j, U at time t, respectively. Then our network transforms into the dynamical system

$$\dot{x} = B(u)\,x - \beta f g^{\mathrm{T}} x\,, \qquad x(0) \geqq \theta\,, \tag{0.1a}$$

$$B(u) = \begin{bmatrix} -k_1 u & k_{-1} & 0 \\ k_1 u & -(k_{-1} + k_2) & k_{-2} u \\ 0 & k_2 & -k_{-2} u \end{bmatrix}, \tag{0.1b}$$

$$f^{\mathrm{T}} = (1, 0, -1)\,, \qquad g^{\mathrm{T}} = (k_3, 0, -k_{-3})\,, \qquad u \geqq 0, \beta \geqq 0 \text{ parameters}\,. \tag{0.1c}$$

It is well known that any solution of (0.1) exists and satisfies $x(t) \geqq \theta$ for all $t \geqq 0$. Furthermore, in the long run, it settles at the steady state $x(u, \beta)$, which is the only solution of the system

$$(B(u) - \beta f g^{\mathrm{T}})\,x = \theta\,, \qquad (\delta, x) = \alpha \geqq 0\,, \tag{0.2}$$

where δ denotes the vector with components $\delta_j = 1$ for all j, and (x, y) is the standard inner product in finite dimensional space. Finally, $\alpha = (\delta, x(0)) \geqq 0$.

We will show that the solution $x(u, \beta)$ can create a *boundary layer phenomenon* at $u = 0$ for small $\beta > 0$, which reflects an instantaneous amplification effect of x for very small $u > 0$. To obtain the boundary layer, the rank of $B(0)$ must be smaller than the rank of $B(u)$ for $u > 0$ (note that rank $(B(0)) = 1 < 2 =$ rank $(B(u))$, $u > 0$, for the matrix (0.1b)).

0.2. In a first approach we freeze the parameters u, β occurring in (0.2) and consider more generally the linear system

$$Bx = b \in R(B), \qquad r = \text{rank}\,(B), \tag{0.3}$$

where $R(B)$ denotes the range of the (N, N)-matrix B. If the eigenvalue 0 of B is *semi-simple* (which means that B is one-to-one on its range), then we can associate with B a regular matrix B^I such that any solution of (0.3) is a linear combination of the (unique) solutions of $N - r + 1$ regular systems $B^I y = q^s$, $s = 1, \ldots, N - r + 1$, with the $q^s \in \mathbb{R}^N$ appropriately chosen. B^I may be regarded as a Lyapunov-Schmidt-reduction of a suitably augmented matrix B. In section 2 we define B^I and prove that B^I is regular if and only if the eigenvalue 0 of B is semi-simple.

In section 3 we replace B by a perturbation $B - C$, and we give a transformation which maps the nullspace $N(B)$ one-to-one onto the nullspace $N(B - C)$ under appropriate conditions on the perturbation C. Furthermore, we can prove a Sherman-Morrison-Woodbury-type formula [5] for the inverse $(B - C)^{-1}$ on $R(B - C)$ in terms of B^{-1} (operating on $R(B)$).

A particular case occurs if C is a rank-1-matrix. This case is considered in section 4 in more detail. Finally, in section 5, we turn to a parameter-dependent problem of the form (0.2) in N dimensions. The assumptions are such that the rank of $B(u)$ is $N - 1$ for $u > 0$ and may drop on the passage $u \to 0$. We give an explicit representation of the solution which reveals the boundary layer effect at $u = 0$ if a rank deficiency indeed occurs at $u = 0$. In that case, the function $x(u, \beta)$ cannot be extended continuously into the origin $u = 0$, $\beta = 0$.

1. Preliminaries

1.1. Let A be an (N, N)-matrix of rank r. Then the following properties are well known:

$$\text{Dim}\,(N(A)) + \text{Dim}\,(R(A)) = N\,; \qquad \text{Dim}\,(R(A)) = r\,; \qquad \text{Dim}\,(N(A)) = \text{Dim}\,(N(A^{\mathsf{T}}))\,. \tag{1.1}$$

Furthermore, any two of the following conditions are equivalent:

$$N(A) \cap R(A) = \{\theta\}\,; \tag{1.2a}$$

$$\mathbb{R}^N = R(A) \oplus N(A)\,; \tag{1.2b}$$

$$A \text{ is one-to-one on } R(A)\,. \tag{1.2c}$$

Finally, *if A is one-to-one on $M \subset \mathbb{R}^N$, then*

$$M = A(M) \subset A(\mathbb{R}^N) = R(A) \tag{1.3}$$

holds.

1.2. Consider vectors $e^1, \ldots, e^s \in \mathbb{R}^N$ and define the set

$$H := \{y \in \mathbb{R}^N\colon (e^j, y) = 0, j = 1, \ldots, s\}\,. \tag{1.4}$$

Then the implication

$$e^j \in N(A^{\mathsf{T}})\,, \qquad j = 1, \ldots, s \Rightarrow R(A) \subset H \tag{1.5}$$

is obvious. Furthermore, $e^1, \ldots, e^s$ *are linearly independent if and only if there are indices* $i_1, \ldots, i_s$ *such that the* (s, s)*-matrix* E *with entries*

$$E_{pq} = e^p_{i_q}\,, \quad p, q = 1, \ldots, s \tag{1.6}$$

is regular. If this is the case, we construct the matrix A^I $\quad (I = \{i_1, \ldots, i_s\} \subset \{1, \ldots, N\})$ with the entries

$$A^I_{pq} = A_{pq} \quad \text{if} \quad p \notin I\,, \quad A^I_{i_p q} = e^p_q \quad \text{if} \quad i_p \in I\,, \qquad q = 1, \ldots, N\,. \tag{1.7}$$

2. Regularization of a singular matrix

2.1. Let B be an (N, N)-matrix and assume

$$B^{\mathsf{T}} e^j = 0\,, \quad j = 1, \ldots, s, e^1, \ldots, e^s\,, \text{ linearly independent, define } H \text{ by (1.4).} \tag{2.1}$$

Then, by 1.2, there is a subset I of $\{1, \ldots, N\}$ such that the corresponding matrix (1.6) is regular. On this basis we construct the matrix B^I with the entries (1.7).

2.2. Any two of the following conditions are equivalent:

B^I *is regular*; (2.2a)

(i) *any system* $B^I x = \delta^j := (0, \dots, 0, 1, 0, \dots, 0)$ (*has j-th component of* δ^j *is* 1) *has a unique solution* h^j *for* $j \in I$, *which forms a basis for* $N(B)$;
(ii) *for* $b \in H$, *define* $b^I \in \mathbb{R}^N$ *by* $b_i^I = b_i$, *if* $i \notin I$, $b_i^I = 0$ *otherwise. Then,* $B^I x = b^I$ *has a unique solution* $z^I \in H$ *and* $Bz^I = b$. *Furthermore,* $R(B) = H$; (2.2b)

$R(B) = H$, B *is one-to-one on* H; (2.2c)

$Bx = \theta$, $(e^j, x) = \alpha_j$, $j = 1, \dots, s$, *has a unique solution for any* $\alpha_j \in \mathbb{R}$, $j = 1, \dots, s$. (2.2d)

Proof: (2.2a) ⇒ (2.2b): The assumption implies the existence of the vectors h^j. Furthermore, these vectors are linearly independent since this is true for the right-hand sides of the equations defining the vectors h^j. Finally, we have $(Bh^j)_i = 0$, $i \notin I, j \in I$ so that

$$\sum_{r=1}^{s} e_{i_r}^p (Bh^j)_{i_r} = (e^p, Bh^j) = (B^T e^p, h^j) = 0, \qquad j \in I, p = 1, \dots, s, \tag{2.3}$$

holds. For any fixed $j \in I$ equation (2.3) is a linear (s, s)-system with the regular matrix (1.6). Therefore, the only solution is the zero element, hence $(Bh^j)_k = 0$ for $k, j \in I$. This establishes $h^j \in N(B), j \in I$, and therefore also Span $(h^{i_1}, \dots, h^{i_s}) \subset N(B)$. In particular, Dim $N(B) \geqq s$ holds. We now turn to (ii) and consider the system $B^I x = b^I$. (2.2a) ensures the existence of the unique solution z^I and by the definition (1.7) we find

$$(Bz^I)_k = b_k \quad \text{for} \quad k \notin I, \qquad (e^p, z^I) = 0 \quad \text{for} \quad p = 1, \dots, s. \tag{2.4}$$

The last set of equalities in (2.4) ensures $z^I \in H$ which, together with the first part, implies

$$\sum_{r=1}^{s} e_{i_r}^p [(Bz^I)_{i_r} - b_{i_r}] = (e^p, Bz^I - b) = (B^T e^p, z^I) = 0 \quad \text{for} \quad p = 1, \dots, s.$$

This is a linear system with the regular matrix (1.6). Therefore, $(Bz^I)_j = b_j$ for $j \in I$ which, together with (2.4), establishes $Bz^I = b$. In particular, $H \subset R(B)$. Recall (1.5) and conclude $H = R(B)$, Dim $R(B) =$ Dim $H = N - s$. Therefore, (1.1a) implies Dim $N(B) = s$, which finishes the proof of (2.2b).

(2.2b) ⇒ (2.2c): The assumption states $N(B) \cap R(B) =$ Span $(h^{i_1}, \dots, h^{i_s}) \cap H = \{\theta\}$ by the construction of the h^j in (2.2b). The implication (2.2c) follows from (1.2).

(2.2c) ⇒ (2.2d): Our assumption together with (1.2) and (1.1), yields

$$N(B) \cap R(B) = \{\theta\}, \qquad \text{Dim } N(B) = s. \tag{2.5}$$

Let $h^{(1)}, \dots, h^{(s)}$ be a basis for $N(B)$. Then the (s, s)-matrix C with entries $C_{ik} = (e^i, h^{(k)})$, $i, k = 1, \dots, s$, is nonsingular. In fact, $Cv = 0$ yields $(e^i, \sum v_k h^{(k)}) = 0$, $i = 1, \dots, s$, so that the element $\sum v_k h^{(k)}$ belongs to the intersection (2.5) and is hence zero implying that all v_k must vanish. Any solution of $Bx = 0$ is of the form $\sum u_k h^{(k)}$ and the second system in (2.2d) requires

$$\sum_{k=1}^{s} (e^j, h^{(k)}) u_k = \alpha_j, \qquad j = 1, \dots, s.$$

This is a linear system for the u's with a regular matrix and has, hence, a unique solution. Therefore, (2.2d) is proved.

(2.2d) ⇒ (2.2a): Assume $B^I z = \theta$, which implies $(Bz)_i = 0$ for $i \notin I$, $(e^j, z) = 0, j = 1, \dots, s$. Moreover,

$$\sum_{r=1}^{s} e_{i_r}^p (Bz)_{i_r} = (e^p, Bz) = (B^T e^p, z) = 0, \qquad p = 1, \dots, s,$$

holds. This is a homogeneous linear system with the regular matrix (1.6) and has, hence, only the trivial solution. As a result, we have $Bz = 0$, $(e^j, z) = 0, j = 1, \dots, s$, and (2.2d) implies $z = 0$. Therefore, $N(B^I) = \{\theta\}$ and (2.2a) follows.

3. Perturbations

3.1. Let B be an (N, N)-matrix satisfying (1.2). To any $x \in R(B)$ there is a unique $Sx \in R(B)$ such that

$$B(Sx) = x, \qquad x, Sx \in R(B), \tag{3.1}$$

holds. S is one-to-one on $R(B)$.

3.2. *Consider two (N, N)-matrices B, C with $R(C) \subset R(B)$ and B satisfying* (1.2). *Let $I - CS$ be one-to-one on $R(C)$. Then $B - C$ is one-to-one on $R(B - C)$ and $R(B) = R(B - C)$.*

Proof: Since $R(C) \subset R(B)$, we have $R(B - C) \subset R(B)$, and $B - C$ maps $R(B)$ into itself. We show that $B - C$ takes $R(B)$ even onto itself: To this end, suppose $(B - C)x = 0$, $x \in R(B)$. Put $y = Bx \in R(B)$ and observe $y = Bx = Cx \in R(C)$. Next, use (3.1) to obtain $x = Sy$. Therefore, $(I - CS)y = Bx - Cx = \theta$, $y \in R(C)$ and, hence, $\theta = y = Bx$, $x \in R(B)$ implying $x = \theta$ by (1.2). This shows that $B - C$ is one-to-one on $R(B)$. Next, (1.3) with $M = R(B)$, $A = B - C$ implies $R(B) \subset R(B - C)$. Since we already know $R(B - C) \subset R(B)$, our proof is complete.

3.3. *Consider the (N, N)-matrices B, C. Let B satisfy* (1.2) *and let $R(C) \subset R(B)$. Then, CS operates on $R(C)$. Assume that $I - CS$ is one-to-one on $R(C)$. Then, by* 3.2, *the matrix $B - C$ is one-to-one on $R(B - C)$ and since $R(B - C)$ can be represented in the form* (1.4) *with vectors $e^1, \dots, e^s \in \mathbb{R}^N$, $s = N - \operatorname{rank}(B - C)$, all statements of* 2.2 *hold for $B - C$. In particular, any solution of the system $(B - C)x = b \in R(B - C)$ has the representation*

$$x = \sum_{j \in I} \alpha_j [(B - C)^I]^{-1} \delta^j + [(B - C)^I]^{-1} b^I, \qquad \alpha_j \in \mathbb{R}, j \in I, \tag{3.2}$$

where I, $(B - C)^I$ and b^I are dependent on $e^1, \dots, e^s$ as defined in 1.2 *and* (2.2b), *respectively.*

Note that (3.2) requires the numerical solution of $s + 1$ linear systems with just one regular matrix $(B - C)^I$ and different inhomogeneities.

Next we are going to represent $N(B - C)$ in terms of $N(B)$:

$$h \in N(B) \Rightarrow (I + S(I - CS)^{-1} C) h \in N(B - C). \tag{3.3}$$

Indeed, we obtain

$$\begin{aligned}(B - C)(I + S(I - CS)^{-1} C) h &= B(S(I - CS)^{-1} Ch) - (I + CS(I - CS)^{-1}) Ch \\ &= (I - CS)^{-1} Ch - (I - CS + CS)(I - CS)^{-1} Ch = \theta\end{aligned}$$

if we apply (3.1) with $x = (I - CS)^{-1} Ch \in R(C) \subset R(B)$. More generally, we claim

$$\begin{aligned}&N(B) = \operatorname{Span}(h^j : j = 1, \dots, s) \\ &\Rightarrow N(B - C) = \operatorname{Span}((I + S(I - CS)^{-1} C) h^j : j = 1, \dots, s).\end{aligned} \tag{3.4}$$

In fact, if $u := \sum \alpha_j h^j$ and $(I + S(I - CS)^{-1} C) u = \theta$, then $\theta = B(I + S(I - CS)^{-1} C) u = B(S(I - CS)^{-1} Cu) = (I - CS)^{-1} Cu$ (note $(I - CS)^{-1} Cu \in R(C) \subset R(B)$ and use (3.1)). Consequently, $\theta = (I + S(I - CS)^{-1} C) u = u = \sum \alpha_j h^j$ and, hence, all $\alpha_j = 0$ since $h^1, \dots, h^s$ is a set of linearly independent vectors. Finally, $R(B - C) = R(B)$ and, therefore,

$$\operatorname{Dim} N(B - C) = N - \operatorname{Dim} R(B - C) = N - \operatorname{Dim} R(B) = \mathit{Dim}\, N(B) = s,$$

which finishes the proof of (3.4).

Our last observation is a generalized Sherman-Morrison-Woodbury formula [5]: *The system*

$$(B - C) y = b \in R(B - C) = R(B)$$

has a unique solution $y \in R(B - C)$ and y has the representation

$$y = S(I + (I - CS)^{-1} CS) b.$$

Indeed, $(I - CS)^{-1} CS(b) \in R(C) \subset R(B)$, hence, $(I + (I - CS)^{-1} CS) b \in R(B) + R(B) \subset R(B)$ and therefore

$$\begin{aligned}&(B - C) S(I + (I - CS)^{-1} CS) b = (I + (I - CS)^{-1} CS) b \\ &- (CS + CS(I - CS)^{-1} CS) b = b + (I - (I - CS) - CS)(I - CS)^{-1} CS(b) = b\end{aligned}$$

which proves

$$(B - C)^{-1} = B^{-1} + B^{-1}(I - CB^{-1})^{-1} CB^{-1} \tag{3.5}$$

if B^{-1}, $(B - C)^{-1}$ mean the existing inverses of B, $B - C$ as operators on $R(B)$, $R(B - C)$, respectively. With this notation, (3.4) implies that the matrix $I + B^{-1}(I - CB^{-1})^{-1} C$ takes $N(B)$ onto $N(B - C)$.

We note in passing that (3.5) coincides with the classical Sherman-Morrison-Woodbury formula [5] if $C = -UV^T$ and if B and $I + V^T B^{-1} U$ are regular. Here, B, U and V are (N, N), (N, k) and (N, k) matrices, respectively.

4. Rank1-perturbations

4.1. A particularly easy situation is a rank1-perturbation $B - fg^{\mathrm{T}}$ of an (N, N)-matrix B satisfying (1.2). *We assume*

$$f, g \in \mathbb{R}^N, \qquad g \neq \theta, \qquad f \in R(B).$$

Then, $R(fg^{\mathrm{T}}) = \operatorname{Span}(f) \subset R(B)$ *(note* $\alpha f = fg^{\mathrm{T}}(\alpha(g, g)^{-1} g)$ *for any* $\alpha \in \mathbb{R}$!). Consequently, the results in 3.3 apply to $B - fg^{\mathrm{T}}$ if $I - fg^{\mathrm{T}}S$ is one-to-one on $R(fg^{\mathrm{T}}) = \operatorname{Span}(f)$. However,

$$(I - fg^{\mathrm{T}}S)(\alpha f) = \alpha f(1 - (g, Sf)), \tag{4.1}$$

so that $I - fg^{\mathrm{T}}S$ *is one-to-one on* Span (f) *if and only if* $(g, Sf) \neq 1$.

4.2. Consider $B - \beta fg^{\mathrm{T}}$ with an (N, N)-matrix B satisfying (1.2) and

$$f, g \in \mathbb{R}^N, \qquad g \neq \theta, \quad f \in R(B), \qquad (g, Sf) < 0, \beta > 0. \tag{4.2}$$

Then, $I - \beta fg^{\mathrm{T}}S$ *is one-to-one on* $R(\beta fg^{\mathrm{T}}) = \operatorname{Span}(f)$ *and all results in* 3.3 *hold for* $B - \beta fg^{\mathrm{T}}$.
Using (4.1), an easy calculation yields

$$(I - \beta fg^{\mathrm{T}}S)^{-1}(\beta fg^{\mathrm{T}})h = \beta(g, h)(I - \beta fg^{\mathrm{T}}S)^{-1} f = \beta(g, h)(1 + \beta|(g, Sf)|)^{-1} f.$$

Therefore, (3.4) shows

$$N(B) = \operatorname{Span}(h^j : j = 1, \ldots, s) \Rightarrow N(B - \beta fg^{\mathrm{T}}) = \operatorname{Span}\left(h^j + \frac{\beta(g, h^j)}{1 + \beta|(g, Sf)|} Sf : j = 1, \ldots, s\right). \tag{4.3}$$

5. Boundary layer phenomena

5.1. Consider (N, N)-matrices $B(u)$ depending on $u \geqq 0$ and satisfying

$$B_{ij}(u) \text{ smooth (at least continuous) for } u \geqq 0, \tag{5.1a}$$

$$B_{ij}(u) \geqq 0, \qquad i \neq j, \qquad B_{ii}(u) = -\sum_{\substack{k=1 \\ k \neq i}}^{N} B_{ki}(u) \quad \text{for} \quad u \geqq 0. \tag{5.1b}$$

Then, $B^{\mathrm{T}}(u)\,\delta = 0$ for $u \geqq 0$. Pick any $j \in \{1, \ldots, N\}$, let $I = \{j\}$ and construct the matrix $B^I(u)$ with entries

$$B^I_{ik}(u) = B_{ik}(u), \qquad i \neq j, \qquad B^I_{jk}(u) = 1, \qquad i, k = 1, \ldots, N. \tag{5.2}$$

This is definition (1.7) with $s = 1$, $e^1 = \delta$. In what follows we assume

$$B^I(u) \text{ regular for } u > 0. \tag{5.3}$$

Then, 2.2 *states*

$$R(B(u)) = \{y \in \mathbb{R}^N : (\delta, y) = 0\}, \qquad u > 0, \tag{5.4a}$$

$$B(u) \textit{ is one-to-one on } R(B(u)), \qquad u > 0, \tag{5.4b}$$

$$N(B(u)) = \operatorname{Span}(h(u)), \qquad u > 0; \tag{5.4c}$$

$$N(B^{\mathrm{T}}(u)) = \operatorname{Span}(\delta), \qquad u > 0. \tag{5.4d}$$

In view of (5.1b) we can choose $h(u)$ in (5.4c) so as to satisfy [1]

$$0 \leqq h_j(u), \quad j = 1, \ldots, N, \qquad (\delta, h(u)) > 0 \quad \text{for} \quad u > 0. \tag{5.5}$$

Next, we define the *rank-1 perturbation* $B(u) - \beta fg^{\mathrm{T}}$ requiring

$$\left.\begin{array}{l} f^{\mathrm{T}} = (1, 0, \ldots, 0, -1) \in \mathbb{R}^N, \qquad g^{\mathrm{T}} = (g_1 \geqq 0, 0, \ldots, 0, g_N \leqq 0) \in \mathbb{R}^N \\ g_1 + |g_N| > 0, \qquad \beta > 0, \qquad (g, S(u) f) < 0, \qquad u > 0, \end{array}\right\} \tag{5.6}$$

where $S(u)$ is defined by (3.1). Then, (4.2) holds and, hence, (4.3) follows. Therefore, *the system*

$$(B(u) - \beta fg^{\mathrm{T}})x = 0, \qquad (\delta, x) = \alpha \geqq 0 \tag{5.7}$$

has the only solution

$$x(u, \beta) = \alpha\left(\frac{h(u)}{(\delta, h(u))} + \beta \frac{(g, h(u))}{(\delta, h(u))} \frac{S(u) f}{1 + \beta|(g, S(u) f)|}\right) \quad \text{for} \quad 0 < u, \beta \geqq 0. \tag{5.8}$$

The special choice of f, g in (5.6) implies that the entries of $B(u) - \beta f g^{\mathrm{T}}$ satisfy (5.1). Therefore, it holds (compare [1] and note $\alpha \geqq 0$)

$$0 \leqq x_j(u, \beta), \quad j = 1, \ldots, N, \quad \text{for} \quad u > 0, \beta \geqq 0. \tag{5.9}$$

5.2. We are interested in the behaviour of $x(u, \beta)$ for $u \to 0$ (note that $x(u, \beta)$ is as yet undefined for $u = 0$). To this end we must study the smoothness of the quantities $h(u)$, $S(u) f$ occurring in the expression (5.8).

(2.2b) *shows that $h(u)$ and $S(u) f$ are the unique solutions of the regularized system*

$$B^I(u) z = \delta^j, \tag{5.10a}$$

$$B^I(u) y = f^I, \tag{5.10b}$$

respectively, where $f_i^I = f_i$ for $i \neq j$, $f_j^I = 0$, $i = 1, \ldots, N$ (compare (2.2b), (ii)). We apply Cramer's rule and Laplace's Theorem, using the *j-th* row of $B^I(u)$ to represent z from (5.10a) as

$$z_i = \varphi_i(u) \Big/ \sum_{k=1}^{N} \varphi_k(u), \quad i = 1, \ldots, N, \tag{5.11}$$

where $\varphi_i(u)$ are appropriate subdeterminants of $B^I(u)$ multiplied by $(-1)^{j+i}$. In particular, (5.1a) guarantees that

$$\varphi_i(u) \text{ is smooth (at least continuous) for } u \geqq 0 \tag{5.12}$$

and our construction of the $\varphi_i(u)$ implies

$$\det(B^I(u)) = \sum_{k=1}^{N} \varphi_k(u) \quad \text{for} \quad u \geqq 0. \tag{5.13}$$

Obviously, the functions $\varphi_i(u)$ depend on j, however, we suppress this dependence in writing.

Next, our assumption (5.1b) enters the scene. It implies that $\operatorname{sign}(\varphi_i(u)) = \operatorname{sign}(\sum \varphi_k(u)) = \operatorname{sign}(\det(B^I(u))$, $i = 1, \ldots, N$. Consequently, we find from (5.13) the equation

$$\sum_{k=1}^{N} |\varphi_k(u)| = \sum \operatorname{sign}(\det(B^I(u))\, \varphi_k(u) = |\det(B^I(u))|. \tag{5.14}$$

From now on we put

$$h_i(u) = |\varphi_i(u)|, \quad i = 1, \ldots, N. \tag{5.15}$$

Then, the following holds,

$$(\delta, h(u)) = |\det(B^I(u))| > 0 \quad \textit{for} \quad u > 0, \tag{5.16a}$$

$$h_i(u) \textit{ is smooth (at least continuous) for } u \geqq 0, \tag{5.16b}$$

$$h_i(u) \geqq 0 \quad \textit{for} \quad u > 0, \tag{5.16c}$$

$$N(B(u)) = \operatorname{Span}(h(u)) \quad \textit{for} \quad u > 0. \tag{5.16d}$$

Next, we turn to (5.10b) and apply Cramer's rule and Laplace's Theorem to find the representation

$$\det(B^I(u))\, y_i = \sum_{k=1}^{N} \Psi_{ik}(u) f_k^I, \quad i = 1, \ldots, N, \tag{5.17}$$

with appropriate subdeterminants $\Psi_{ik}(u)$ of $B^I(u)$ multiplied by $(-1)^{i+k}$. Therefore, $\Psi_{ik}(u)$ is smooth (at least continuous) for $u \geqq 0$. Using (5.16 a) and the matrix $\sigma(u)$ with elements $\sigma_{ik}(u) = \operatorname{sign}(\det(B^I(u)))\, \Psi_{ik}(u)$ if $k \neq j$ and $\sigma_{ij}(u) = 0$, $i, k = 1, \ldots, N$, equation (5.17) turns into $(\delta, h(u))\, y = \sigma(u) f$.

By (5.3) we may consider $\det(B^I(u))$ of one sign for $u \geqq 0$. Hence,

$$\sigma_{ik}(u) \text{ are smooth (at least continuous) for } u \geqq 0. \tag{5.18}$$

Since $y = S(u) f$, we finally obtain

$$(\delta, h(u))\, S(u) f = \sigma(u) f, \tag{5.19}$$

and (5.8) shows

$$x(u, \beta) = \alpha \left\{ \frac{h(u)}{(\delta, h(u))} + \frac{(g, h(u))}{(\delta, h(u))} \frac{\beta \sigma(u) f}{(\delta, h(u)) + \beta |(g, \sigma(u) f)|} \right\} \tag{5.20}$$

with all entries smooth for $u \geqq 0$, $\beta \geqq 0$ as stated in (5.16b) and (5.18). Consequently, the function (5.20) is smooth if $u > 0$ for any $\beta > 0$. If $\det(B^I(0)) \neq 0$, then (5.16a) implies $(\delta, h(0)) > 0$ and, hence, (5.20) is defined at the origin ($u = 0$,

$\beta = 0$). Moreover, it is a continuous function for $u \geqq 0, \beta \geqq 0$ if we define $x(0, 0) = \alpha h(0) (\delta, h(0))^{-1}$. In particular, we have

$$x(u, \beta) \approx x(0, 0) = \alpha \frac{h(0)}{(\delta, h(0))} \quad \text{for} \quad u \in [0, u_0], \beta \approx 0 \ (u_0 > 0 \text{ arbitrary but fixed}), \tag{5.21}$$

so that we may consider (5.7) to be a regular perturbation of

$$B(u) x = 0, \qquad (\delta, x) = \alpha \tag{5.22}$$

(which has the unique solution $x(u, 0)$) if $\beta > 0$ is our parameter defining the perturbation.

5.3. The situation is different if $\det(B^I(0)) = 0$, which means that $(\delta, h(0)) = |\det(B^I(0))| = 0$ (see (5.16a)) so that $\operatorname{rank}(B(0)) < N - 1 = \operatorname{rank}(B(u))$, $u > 0$, by (5.4a) and (1.1). In the first place, (5.16b) and (5.18) imply that $h(u)$ and $\sigma(u)f$ are continuous if $u \geqq 0$. Moreover, $0 \leqq h_j(u) \leqq (\delta, h(u)) = |\det(B^I(u))|$, $j = 1, \ldots, N$, follows from (5.16a, c), showing that $h_j(0) = 0$, $j = 1, \ldots, N$, $(\delta, h(0)) = 0$ holds. *Assuming*

$$h_j(u) = u h_j^0(u),\ h_j^0 \text{ continuous for } u \geqq 0, \qquad j = 1, \ldots, N, \qquad (\delta, h^0(0)) > 0 \tag{5.23}$$

the limit $h(u) (\delta, h(u))^{-1} \to h^0(0) (\delta, h^0(0))^{-1}$ *as* $u \to 0$ *exists.*

Finally, we need the assumption

$$(g, \sigma(0) f) \neq 0 \tag{5.24}$$

to obtain the limits

$$x(u, \beta) \to \alpha \left\{ \frac{h^0(0)}{(\delta, h^0(0))} + \frac{(g, h^0(0))}{(\delta, h^0(0))} \frac{\sigma(0) f}{|(g, \sigma(0) f)|} \right\}, \qquad \beta > 0, \tag{5.25a}$$

$$x(u, 0) \to \alpha \frac{h^0(0)}{(\delta, h^0(0))}, \quad \text{as} \quad u \to 0. \tag{5.25b}$$

Note that $\sigma(0) f \neq 0$ follows from (5.24). Therefore, the last term in the parentheses in (5.25a) vanishes if and only if $(g, h^0(0)) = 0$. Only then can (5.20) be defined at the origin $x(0, 0) = \alpha h^0(0) (\delta, h^0(0))^{-1}$ to turn $x(u, \beta)$ into a continuous function for $u \geqq 0$, $\beta \geqq 0$.

The more interesting case occurs if $\det(B^I(0)) = 0$, and if (5.23) as well as

$$(g, \sigma(0) f) \neq 0, \qquad (g, h^0(0)) \neq 0 \tag{5.26}$$

hold. *Then, for any* $\beta \geqq 0$ (arbitrary but fixed), *we can define* $x(0, \beta)$ *such that the resulting functions*

$$x(u, \beta) = \alpha \left\{ \frac{h^0(u)}{(\delta, h^0(u))} + \frac{(g, h^0(u))}{(\delta, h^0(u))} \varphi(u, \beta) \right\}, \qquad 0 \leqq u, \qquad 0 \leqq \beta, \tag{5.27a}$$

$$\varphi(u, \beta) := \begin{cases} \dfrac{\beta \sigma(u) f}{(\delta, h(u)) + \beta |(g, \sigma(u) f)|} & \text{for} \quad 0 \leqq u, \beta > 0 \\ 0 & \text{for} \quad 0 \leqq u, \beta = 0 \end{cases} \tag{5.27b}$$

are continuous for any fixed $\beta \geqq 0$. *In particular, we obtain*

$$\varphi(0, \beta) = \begin{cases} \dfrac{\sigma(0) f}{|(g, \sigma(0) f)|} & \text{for} \quad \beta > 0 \\ 0 & \text{for} \quad \beta = 0 \end{cases}$$

and, in a boundary layer, $0 < u = O(\beta)$,

$$\varphi(u, \beta) \approx \frac{\sigma(u) f}{|(g, \sigma(u) f)|} \quad \text{if} \quad 0 < u(\delta, h^0(u)) \ll \beta |(g, \sigma(u) f)|$$

holds. Note that $x(u, \beta)$ is no longer continuous at the origin ($u = 0$, $\beta = 0$).

As an example, let us take a closer look at the system (0.2) with the entries (0.1b, c). Then we find

$$h(u) = u(k_{-1} k_{-2}, k_1 k_{-2} u, k_1 k_2) =: u h^0(u),$$

$$\sigma(u) f = (-(k_{-2} u + k_{-1} + k_2), u(k_{-2} - k_1), k_1 u + k_{-1} + k_2)$$

and, therefore,

$$(\delta, h^0(0)) = k_{-1} k_{-2} + k_1 k_2 > 0, \qquad (g, \sigma(0) f) = -(k_3 + k_{-3})(k_2 + k_{-1}) < 0,$$

$$(g, h^0(0)) = k_3 k_{-1} k_{-2} - k_{-3} k_1 k_2.$$

Hence, (5.23) and (5.26) are satisfied if

$$k_1 k_2 k_{-3} \neq k_{-1} k_{-2} k_3 .$$

Since $\det(B^1(0)) = 0$, the boundary layer phenomenon occurs. From (5.20) we learn

$$x_1(u, \beta) = \frac{\alpha}{(\delta, h^0(u))} \left\{ k_{-1}k_{-2} - \beta(k_3 k_{-1} k_{-2} - k_{-3} k_1 k_2) \frac{k_{-2}u + k_{-1} + k_2}{(\delta, h(u)) + \beta |(g, \sigma(u) f)|} \right\}, \tag{5.28}$$

and it shows that the chemical X_1 experiences an amplification for small stimuli $u > 0$ if

$$\beta > 0 , \qquad k_3 k_{-1} k_{-2} > k_{-3} k_1 k_2 .$$

6. Numerical considerations

6.1. Consider a system

$$(B(u) - \beta f g^{\mathrm{T}}) x = 0 , \qquad (\delta, x) = \alpha \geqq 0 \tag{6.1}$$

as in 5.1 under the assumptions (5.1), (5.3) and (5.6). The representation (5.8) of the unique solution $x(u, \beta)$ allows the discussion of its qualitative behaviour. For the actual computation of $x(u, \beta)$ it is more appropriate to return to (3.2) and to solve

$$(B(u) - \beta f g^{\mathrm{T}})^I x = \alpha \delta^j \qquad (I = \{j\}, j \in \{1, \dots, N\} \text{ arbitrary but fixed}) \tag{6.2}$$

with standard routines for linear systems. Note that

$$(B(u) - \beta f g^{\mathrm{T}})^I \text{ is regular for } 0 < u \leqq u_0, \beta \geqq 0 \tag{6.3}$$

(compare 5.1, 4.2 and 3.3). If, in addition, $(B(0) - \beta f g^{\mathrm{T}})^I$ ($\beta > 0$) is regular, then (6.1) has a unique solution for any $u \geqq 0$, $\beta > 0$ which can numerically be obtained via (6.2).

6.2. Let $B(u)$, f, g be given by (0.1 b, c). Then,

$$\operatorname{rank}(B(0)) = 1 < 2 = \operatorname{rank}(B(u)) , \qquad 0 < u ,$$
$$\text{however, } \operatorname{rank}(B(0) - \beta f g^{\mathrm{T}}) = 2 , \qquad 0 < \beta .$$

Therefore, (6.3) holds for $u \geqq 0$ allowing for the application of standard routines to solve (6.2) yet observing the boundary layer phenomenon near $u = 0$ if $\beta > 0$ but small.

We can study this, taking, for example,

$$k_1 = 10 , \quad k_2 = 0.1 , \quad k_3 = 3 , \quad k_{-j} = 1 , \quad j = 1, 2, 3 , \quad \alpha = 1 . \tag{6.4}$$

The result is given in figure 2 which shows the boundary layer building up for decreasing β and the approach

$$x_1(u, \beta) \to \alpha \frac{h_1^0(u)}{(\delta, h^0(u))} = \frac{\alpha k_{-1} k_{-2}}{k_1 k_{-2} u + k_1 k_2 + k_{-1} k_{-2}} = \frac{1}{10u + 2} \quad \text{as } \beta \to 0, u > 0 \text{ fixed} ,$$

which follows from (5.27) and (5.28). All calculations we have performed using the regularization (6.2) for the example above and in a series of other examples (with random numbers for the rate constants) showed a very stable performance in solving the equations (6.2).

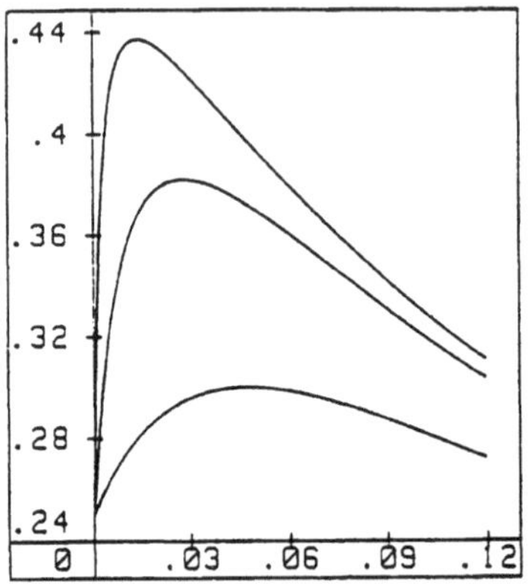

Fig. 2. The solutions $x_1(u, \beta)$ of (0.2) for $\beta = 0.3$, .005, .001 (bottom curve to top curve) with the other constants given in (6.4). Horizontal axis = u, vertical axis = x_1.

References

[1] BOHL, E.: Monotonie: Lösbarkeit und Numerik bei Operatorgleichungen. Springer Tracts in Natural Philosophy, Bd. 25, Springer-Verlag, Berlin 1974.
[2] BOHL, E.: Activator actions initiating the cascade of vision. To appear in: STREHMEL, K. (ed.): Proc. 5th Seminar NUMDIFF-5, Halle 1989. Teubner-Texte zur Mathematik, Leipzig 1990.
[3] BOHL, E.: Structural amplification in chemical networks. Submitted for publication.
[4] GOLDBETER, A.: Models for oszillations and excitabilty in biochemical systems. In SEGEL, L. A. (ed.): Mathematical models in molecular and cellular biology. Cambridge University Press, Cambridge 1980, pp. 248–291.
[5] GOLUB, G.; VAN LOAN, C.: Matrix computations. North Oxford Academic Publishing Co., Oxford 1986.
[6] MARTIEL, J.-L.; GOLDBETER, A.: A model based on receptor desensitization for cyclic AMP signaling in Dictyostelium cells. Biophys. J. **52** (1987), 807–828.

Received May 2, 1990

Address: Prof. Dr. ERICH BOHL, Fakultät für Mathematik, Universität Konstanz, Postfach 5560, W-7750 Konstanz, Germany

BOOK REVIEWS

Lawden, D. F., Elliptic Functions and Applications. Berlin etc., Springer-Verlag 1989. XIV, 334 pp., 23 figs., DM 124,00. ISBN 3-540-96965-9 (Applied Mathematical Sciences 80)

Die Theorie der elliptischen Funktionen wurde von den großen Mathematikern des vergangenen Jahrhunderts begründet, vor allem von Abel, Jacobi und Weierstraß als „a treasurehouse of results whose variety, aesthetic appeal, and capacity for arousing our astonishment have not since been equaled by research in any other area.“ Aus dem heutigen Universitätsstudium ist diese Theorie sehr zu unrecht fast völlig verschwunden, so daß Verf. es sich zur Aufgabe gestellt hat, sie aus ihrem Dornröschenschlaf zu erwecken.

Der Einstieg in die Theorie wurde über die Thetafunktionen gewählt, die als spezielle Lösungen der Wärmeleitungsgleichung definiert werden. Die Methode der Trennung der Veränderlichen führt hiervon unmittelbar zu ihrer Darstellung durch einfache Potenzreihen, aus denen dann in bequemer Weise alle ihre wichtigen Eigenschaften hergeleitet werden. Nach der Definition der Jacobischen elliptischen Funktionen als Quotienten von Thetafunktionen ergibt sich dann sofort deren ganzer Formelapparat, insbesondere deren Additionstheoreme und Differentialgleichungen. Durch Umkehrung der elliptischen Funktionen wird anschließend die Theorie der elliptischen Integrale entwickelt. Die nächsten beiden Kapitel befassen sich mit unterschiedlichen geometrischen Anwendungen (auf Ellipsen, Ellipsoide, Lemniskaten, Spiralen und Ovale) bzw. physikalischen Anwendungen (auf Schwingungen unterschiedlicher Art, relativistische Planetenbewegung, wirbelnde Kette, freie Rotation, Stromfluß in einer rechteckigen Leiterplatte, Plattenkondensator).

Während die bisherigen Ausführungen kaum mehr als Grundkenntnisse der Analysis erfordern, werden in den letzten vier Kapiteln tieferliegende funktionentheoretische Hilfsmittel herangezogen, um die gewonnenen Einzelergebnisse zu einer allgemeinen Theorie abzurunden, die bis zu den Modultransformationen reicht. Hier wird auch die Theorie der Weierstraßschen elliptischen Funktionen mit dem Einstieg über die Sigmafunktionen geboten, ihr Zusammenhang mit den vorhergehenden Kapiteln dargestellt sowie eine weitere Reihe neuer konkreter Anwendungen behandelt.

Alle Kapitel sind mit zahlreichen Übungsaufgaben versehen, so daß das gut geschriebene Buch auch zum Selbststudium bestens geeignet ist, obwohl mathematische Sätze nirgends expilizit formuliert werden. Es folgt noch ein Anhang mit zwei kurzen analytischen Teilen, zwei Basic-Programmen und auf 60 Seiten vor allem 13 sehr nützlichen Tabellen über die behandelten Funktionen.

Vergeblich sucht man im Stichwortverzeichnis nach einem Hinweis auf die Definition der elliptischen Funktionen, doch findet man etwas versteckt auf S. 27, daß es sich dabei um bis auf Pole holomorphe doppeltperiodische Funktionen handelt.

Rostock L. Berg

Mercier, B., An Introduction to the Numerical Analysis of Spectral Methods. Berlin etc., Springer-Verlag 1989. V, 154 pp., DM 39,–. ISBN 3-540-51106-7 (Lecture Notes in Physics 318)

Das vorliegende Buch ist die englische Übersetzung eines Reports, der 1981 von der französischen Atomenergiekommission veröffentlicht wurde und auf eine von Ciarlet und Raviart veranlaßte Vorlesungsreihe zurückgeht. Spektralmethoden sind Näherungsverfahren zur Lösung partieller Differentialgleichungen. Die Näherungslösung wird dabei durch Entwicklungen nach Eigenfunktionen geeigneter Differentialoperatoren dargestellt. Hierher gehören Projektionsverfahren (Galerkinverfahren) und Kollokationsverfahren mit speziellen Knotensystemem. Behandelt werden insbesondere Projektionsverfahren in Sobolewräumen über einem eindimensionalen Intervall mit dem trigonometrischen System der Basisfunktionen bei periodischen Randbedingungen. Kollokationsverfahren mit den zugehörigen Interpolationen werden für äquidistante Knotensysteme und das System der Chebyshev-Knoten dargestellt. Diese Verfahren werden auf einfache lineare partielle Differentialgleichungen 1. Ordnung und lineare Wärmeleitungsprobleme in einer Raumdimension angewandt. Dabei werden Konvergenzaussagen der üblichen Art und Stabilitätsaussagen für die zeitdiskretisierten Gleichungen ausführlich bewiesen. Es wird angedeutet, daß bei der Lösung der zeitdiskretisierten Gleichungen die schnelle Fourier-Transformation (FFT) eine Rolle spielen kann. Bedauerlich ist, daß das weite Feld der nichtlinearen Gleichungen, wo die Spektralmethoden im Zusammenwirken mit der FFT so wichtig geworden sind, völlig unerwähnt bleibt. Die über das Jahr 1981 hinausführende Literatur ist nur mit einem Zitat vertreten. Als eine erste und elementare Einführung in das Gebiet der Spektralmethoden kann das vorliegende Buch dennoch empfohlen werden.

Berlin K. Zacharias

Rodrigues, J. F. (ed.), Mathematical Models for Phase Change Problems. Proceedings European Workshop, Obidos, Portugal, October 1–3, 1988. Basel etc., Birkhäuser Verlag 1989. 408 pp., sfr 118.–. ISBN 3-7643-2309-4 (International Series of Numerical Mathematics 88)

Unter der Federführung des Instituts für Mathematik (Prof. Dr. K.-H. Hoffmann) der Universität Augsburg arbeiten seit August 1986 sechs Mathematik-Institute europäischer Universitäten auf dem Gebiet „Mathematical Models of Phase Transitions and Numerical Simulation“ eng zusammen. Dieser Vertrag steht unter EG-Registration und vereint alle größeren mathematischen Forschungskapazitäten dieses Arbeitsfeldes im westeuropäischen Raum (und darüber hinaus). Aus dem zusammenfassenden Report (K.-H. Hoffmann) geht u. a. hervor,

– daß der vorgestellte Workshop bereits die 8. Zusammenkunft der teilnehmenden Seiten ist,

– daß gemeinsame Forschungsprojekte verfolgt wurden, die zu theoretischen Konzepten, effektiven Techniken und neueröffneten Forschungsrichtungen führten,

– daß viele Publikationen angeregt wurden (eine Auswahl von 104 Arbeiten des Zeitraumes 86–89 ist enthalten).

Zum Workshop wurden je 6–7 Beiträge zu folgenden 3 Unterthemen vorgestellt:

1. Verallgemeinerte Phasenübergänge, 2. Stefan-Aufgaben, 3. Vermischte Probleme.

Behandelt werden Fragen zur Konstruktion mathematischer Modelle (u. a. Thermodynamik der Phasenübergänge, Erhaltungssätze der Kontinuumsmechanik), zur Beschreibung besonderer physikalischer Effekte (wie Mischphasengebiete, Unterkühlung, Einfluß der Oberflächenspannung auf den Phasenübergang), zur Existenz von Lösungen der mathematischen Aufgaben und zur Technik der numerischen Lösung der Probleme (insbesondere zur adaptiven Netzverfeinerung).

Chemnitz W. Weinelt

Claßen, L.; Wiesner, U., Wissensspeicher 80286-Programmierung. Berlin, Verlag Technik 1990. 240 S., DM 29,80. ISBN 3-341-00867-5 (Technische Informatik)

Der vorliegende Band behandelt den Aufbau, die interne Struktur, die verschiedenen Betriebsarten und natürlich den ganzen Komplex der Programmierung des 80286 aus dem Blickwinkel der Software. Der Band ist gut zur Einarbeitung in die Konzepte des 80286 geeignet und für die Aus- und Weiterbildung an den Universitäten, Hoch- und Fachschulen verwendbar. Dabei werden Grundkenntnisse der Programmierung vorausgesetzt. Für den erfahrenen Programmierer soll der Band den Charakter eines Nachschlagewerkes mit hoher Informationsdichte haben.

(Aus dem Vorwort)

ZAMM · Z. angew. Math. Mech. 71 (1991) 7/8, 233–239 Akademie-Verlag

KUHNERT, F.

Über die Sensibilität von Eigenwerten und Eigenvektoren

Dem Gedenken an LOTHAR COLLATZ gewidmet

Betrachtet werden allgemeine Eigenwertaufgaben mit parameterabhängigen Matrizen. Sensibilitätsanalyse bedeutet Untersuchung des Einflusses dieser Parameter auf die Eigendaten. Für die Eigenwerte werden lineare und quadratische Glieder angegeben. Untersucht werden auch iterative Verfahren für inverse Aufgaben, die entstehen, wenn einzelne Eigenwerte vorgegeben werden und die entsprechenden Parameter zu identifizieren sind.

The general eigenvalue problem with matrices depending on parameters will be considered. Sensibility analysis means the examination of the influence of the parameters to eigenvalues and eigenvectors. Linear and quadratic parts in series of eigenvalues are obtained. In addition inverse problems are considered, which result from parameter identification provided that some eigenvalues are given.

Рассматриваются общие проблемы на собственные значения с матрицами, зависящимися от параметров. Анализ чувствительности означает исследование влияния этих параметров на собственные значения и векторы. Для собственных значений получены линейные и квадратичные слагаемые. Рассматриваются также обратные задачи, возникающие в проблеме идентификации параметров при заданных собственных значений.

MSC (1980): 65F15

1. Einleitung

Die Untersuchung des Schwingungsverhaltens von mechanischen Systemen kann über den Weg einer immer feineren mechanischen Modellierung (siehe z. B. [2] und [7] für den Fall rotierender Maschinen) mehr und mehr in die Entwurfsphase dieser Systeme verlegt werden und ermöglicht die Betrachtung einer Vielzahl von Varianten dieses Systems. Von besonderer Bedeutung ist dabei die Analyse des Einflusses von Systemparametern auf das Schwingungsverhalten, d. h. die Analyse des funktionellen Zusammenhangs von Systemparametern und Eigendaten des Systems (dazu seien vor allem Eigenwerte und Eigenvektoren, aber auch Eigenunterräume gezählt). Wir gehen dabei von der Voraussetzung aus, daß die mathematische Formulierung des Eigenschwingungszustands über die mechanische Modellierung auf eine allgemeine Eigenwertaufgabe der Form

$$AU - \lambda BU = 0 \tag{1}$$

mit quadratischen Matrizen A und B und einspaltigen Matrizen $U \neq 0$ führt. Mit λ werden die (i. a. komplexwertigen) Eigenwerte bezeichnet.

Es wird vorausgesetzt, daß die Matrizen A und B von Parametern $p_1, \ldots, p_s$ abhängen, $p = (p_1, \ldots, p_s)^{\mathrm{T}} \in P \subset \mathbb{R}^s$ ($\mathbb{R}^s$ ist der Euklidische s-dimensionale Raum), so daß auch die Eigenwerte λ und Eigenvektoren U als Funktionen des Parametervektors p aufzufassen sind:

$$\lambda = \lambda(p), \qquad U = U(p). \tag{2}$$

In der Arbeit werden Formeln für die linearen und quatratischen Anteile in den Reihenentwicklungen des Eigenwerts $\lambda(p)$ nach den Parametern p bereitgestellt. Damit wird der globale Charakter der Abhängigkeit der Eigenwerte vom Parametervektor p hervorgehoben und die Berechnung der sogenannten Sensitivitäten (siehe z. B. [7]) verallgemeinert, die Hinweise auf den Einfluß jedes einzelnen Parameters p_k auf den Eigenwert $\lambda(\ldots, p_k, \ldots)$ geben.

Der Herleitung von Formeln zur Berechnung linearer Anteile der Eigenvektoren erscheint gerechtfertigt, wenn man sie z. B. als Startvektoren für Vektoriterationen verwendet. In dieser Arbeit dienen sie als Hilfsmittel zur Berechnung der quadratischen Korrekturanteile der Eigenwerte.

2. Mathematische Modellierung

Die mathematische Modellierung des Eigenschwingungsverhaltens von mechanischen Systemen führt unter Berücksichtigung von Dämpfungsgliedern auf die quadratische Eigenwertaufgabe

$$(\lambda^2 M + \lambda D + C) X = 0. \tag{3}$$

Hier bezeichnen M, D und C die quadratischen n-zeiligen Trägheits-, Dämpfungs- bzw. Steifigkeitsmatrizen und $X \in \mathbb{C}^n$ (Raum der komplexwertigen Vektoren) den zum Eigenwert λ gehörenden *(Rechts)-Eigenvektor*, $X \neq 0$. In gleicher Weise wird durch

$$Y^*(\lambda^2 M + \lambda D + C) = 0 \tag{4}$$

der Begriff des *Links-Eigenvektors* $Y = (y_1, \ldots, y_n)^T$ eingeführt (es bedeutet Y^* der zu $Y \in \mathbb{C}^n$ adjungierte zeilige Vektor $(\bar{y}_1, \ldots, \bar{y}_n)$).

Konstruiert man die quadratischen $(2n)$-zeiligen Blockmatrizen

$$A = \begin{bmatrix} 0 & I \\ \hline -C & -D \end{bmatrix}, \qquad B = \begin{bmatrix} I & 0 \\ \hline 0 & M \end{bmatrix}, \tag{5}$$

so kann man das quadratische Eigenwertproblem (3) in das lineare Eigenwertproblem (1) überführen, wobei die Rechts- bzw. Links-Eigenvektoren U und V der Aufgabe (1) durch

$$U = \begin{bmatrix} X \\ \hline \lambda X \end{bmatrix}, \qquad V = \begin{bmatrix} -\dfrac{1}{\bar{\lambda}} CY \\ \hline Y \end{bmatrix} \tag{6}$$

festgelegt sind. Dabei setzen wir voraus, daß die Steifigkeitsmatrix C symmetrisch und positiv definit ist, andernfalls müßte in der Formel für V die Matrix C gegen C^* ausgetauscht werden. Es ist folglich ausreichend, Aufgaben der Form (1) zu untersuchen.

Es sei $A = A(p) = A(p_1, \ldots, p_s)$ und $B = B(p) = B(p_1, \ldots, p_s)$. Wir setzen voraus, daß diese Matrix-Funktionen nach allen Parametern p_k differenzierbar sind und daß die zweiten Ableitungen $\partial^2 A/\partial p_k\, \partial p_j$ und $\partial^2 B/\partial p_k\, \partial p_j$ vernachlässigbar klein werden. Das ist z. B. der Fall, wenn A und B lineare Funktionen in diesen Parametern sind:

$$A = A_0 + p_1 A_1 + \ldots + p_s A_s, \qquad B = B_0 + p_1 B_1 + \ldots + p_s B_s. \tag{7}$$

Hier bezeichnen $A_0, \ldots, A_s, B_0, \ldots, B_s$ konstante quadratische n-zeilige Matrizen.

Um Ableitungen der Eigenwerte λ und Eigenvektoren U nach den Parametern p_k zu erhalten, bilden wir in der Beziehung (1) formal Ableitungen nach p_k bzw. nach p_k und p_j. Nach einmaliger Differentiation ergibt sich

$$\left(\frac{\partial A}{\partial p_k} - \lambda \frac{\partial B}{\partial p_k}\right) U - \frac{\partial \lambda}{\partial p_k} BU + (A - \lambda B) \frac{\partial U}{\partial p_k} = 0. \tag{8}$$

Differenzieren wir diese Beziehung erneut und lassen alle höheren Ableitungen der Matrizen A und B weg, so folgt

$$\begin{aligned} &\left(\frac{\partial A}{\partial p_k} - \lambda \frac{\partial B}{\partial p_k}\right) \frac{\partial U}{\partial p_j} + \left(\frac{\partial A}{\partial p_j} - \lambda \frac{\partial B}{\partial p_j}\right) \frac{\partial U}{\partial p_k} - \left(\frac{\partial \lambda}{\partial p_k} \frac{\partial B}{\partial p_j} + \frac{\partial \lambda}{\partial p_j} \frac{\partial B}{\partial p_k}\right) U \\ &- B \left(\frac{\partial \lambda}{\partial p_k} \frac{\partial U}{\partial p_j} + \frac{\partial \lambda}{\partial p_j} \frac{\partial U}{\partial p_k}\right) - \frac{\partial^2 \lambda}{\partial p_k\, \partial p_j} BU + (A - \lambda B) \frac{\partial^2 U}{\partial p_k\, \partial p_j} = 0, \qquad k, j = 1, \ldots, s. \end{aligned} \tag{9}$$

Wir vermerken, daß die Ableitungen der Eigendaten λ und U nicht für alle Parameterwerte existieren müssen, wie das folgende Beispiel zeigt. Es sei

$$A = \begin{bmatrix} 0 & 0 & 0 \\ 0 & 0 & 0 \\ 0 & 0 & 2 \end{bmatrix} + p_1 \begin{bmatrix} 1 & 0 & 0 \\ 0 & -1 & 0 \\ 0 & 0 & 1 \end{bmatrix} + p_2 \begin{bmatrix} 0 & 1 & 0 \\ 1 & 0 & 0 \\ 0 & 0 & 0 \end{bmatrix}, \qquad B = I.$$

Dann ergeben sich die Eigenwerte zu $\lambda_{1,2} = \mp (p_1^2 + p_2^2)^{1/2}$ und $\lambda_3(p) = 2 + p_1$. Für die Parameterwerte $p_1 = p_2 = 0$ existiert für die Eigenwerte λ_1 und λ_2 keine Ableitung.

Durch Multiplikation von links mit dem Eigenvektor V^* gewinnt man aus Formel (8) infolge $V^*(A - \lambda B) = 0$ die Beziehung

$$\frac{\partial \lambda}{\partial p_k} = (V^* BU)^{-1}\, V^* \left(\frac{\partial A}{\partial p_k} - \lambda \frac{\partial B}{\partial p_k}\right) U, \qquad k = 1, \ldots, s.$$

Wir halten die Parameter $(p_1^0, \ldots, p_s^0)^T = p^0$ fest und setzen in Formel (9) $p = p^0$. Multiplizieren wir dann mit $\Delta p_k = p_k - p_k^0$ und summieren über k, so erhalten wir für die Differenz $\lambda(p) - \lambda(p^0)$ den folgenden linearen Korrekturterm $\Delta\lambda$:

$$\Delta\lambda = \sum_{k=1}^{s} \Delta p_k \left.\frac{\partial \lambda}{\partial p_k}\right|_{p=p^0} = (V^{0*} B^0 U^0)^{-1}\, V^{0*} \sum_{k=1}^{s} \left.\left(\Delta p_k \frac{\partial A}{\partial p_k} - \lambda^0\, \Delta p_k \frac{\partial B}{\partial p_k}\right)\right|_{p=p^0} U^0 \tag{10}$$

(hierbei bedeutet der obere Index „0", daß in der jeweiligen Größe $p = p^0$ zu setzen ist, z. B. $\lambda^0 = \lambda(p^0)$, $B^0 = B(p^0)$). Die in dieser Formel stehende Summe

$$\sum_{k=1}^{s} \left(\Delta p_k \frac{\partial A}{\partial p_k} - \lambda^0 \Delta p_k \frac{\partial B}{\partial p_k} \right) \bigg|_{p=p^0}$$

ist gerade die Linearisierung der Matrix

$$(A(p) - \lambda^0 B(p)) - (A(p^0) - \lambda^0 B(p^0)) .$$

Ersetzen wir in Formel (10) die Summe durch diese Differenz (das ist gerechtfertigt, wenn wir nur an linearen Korrekturgliedern interessiert sind) und beachten

$$(A(p^0) - \lambda^0 B(p^0)) U^0 = 0 ,$$

so ergibt sich als linearer Korrekturterm (in Bezug auf $\Delta p = p - p^0$) für den Eigenwert $\lambda(p)$ der Ausdruck

$$\Delta\lambda = (V^{0*} B^0 U^0)^{-1} V^{0*} (A(p) - \lambda^0 B(p)) U^0 . \tag{11}$$

Dann können wir

$$\lambda(p) \approx \lambda(p^0) + \Delta\lambda \tag{12}$$

setzen.

Im Falle der linearen Abhängigkeit der Matrizen $A(p)$ und $B(p)$ von den Parametern $p_1, \ldots, p_s$ sind die Formeln (10) und (11) offenbar identisch.

Formel (11) zeigt, daß für die Korrektur des Eigenwertes $\lambda(p^0)$ nur die Kenntnis dieses Eigenwertes und der zugehörenden Links- und Rechts-Eigenvektoren $V(p^0)$ und $U(p^0)$ der Aufgabe (1) erforderlich ist (d. h. andere Eigenwerte und Eigenvektoren werden für diese Korrektur nicht benötigt). Weiter kann vermerkt werden, daß Formel (11) (im Gegensatz zu (10)) ohne die Berechnung von Ableitungen auskommt. Setzt man in Formel (10) $p_k = 1$ und $p_j = 0$ $(j \neq k)$, so ergeben sich die bekannten Formeln für die Sensitivitäten (siehe [7]).

Mit Hilfe der Formeln (11) und (12) können die Veränderungen der Eigenwerte in Abhängigkeit vom Parametervektor p sehr gut beobachtet werden. Es können auch einzelne oder mehrere Eigenwerte $\lambda(p)$ „relativ" stabilisiert werden, indem man fordert, daß ihre linearen Korrekturterme verschwinden. Setzt man in Formel (11) $\Delta\lambda = 0$, so ergibt sich eine Gleichung für den Vektor p. Verwendet man Formel (10), so ist diese Gleichung linear in den Parametern p_k. Betrachtet man gleichzeitig mehrere Eigenwerte, so ergibt sich ein System linearer Gleichungen für $p_1, \ldots, p_s$. Die Lösungen p dieses Systems ergeben für die Differenzen $\lambda(p) - \lambda(p^0)$ eine Größenordnung $O(\| p - p^0 \|^2)$.

Analog zu (12) kann man auch für die Eigenvektoren $U = U(p)$ und $V = V(p)$ lineare Korrekturanteile ΔU und ΔV ermitteln. In der Tat, aus Formel (8) ergibt sich z. B. das lineare Gleichungssystem

$$(A^0 - \lambda^0 B^0) \Delta U = \Delta\lambda B^0 U^0 - (A(p) - \lambda^0 B(p)) U^0 \tag{13}$$

zur Bestimmung des Vektors ΔU. Die Systemmatrix $A^0 - \lambda^0 B^0$ ist offenbar singulär, die rechte Seite liegt aber infolge (11) orthogonal zum Links-Eigenvektor V^0, so daß (13) sinnvoll und ΔU als Normallösung des singulären Systems (13) zu interpretieren ist. Bemerkenswert ist, daß lediglich die rechte Seite des Systems (13) vom Parametervektor p abhängt, nicht aber die Systemmatrix, d. h. bei Bedarf können in Abhängigkeit von p mehrere Varianten für ΔU bei gleichbleibender Systemmatrix ermittelt werden.

Unter Verwendung von Formel (9) ist es möglich, neben dem linearen Korrekturglied nach Formel (10) bzw. (11) für die Differenz $\lambda(p) - \lambda(p^0)$ einen quadratischen Korrekturanteil zu berechnen. In der Tat, multipliziert man die Beziehung (9) von links mit V^*, so wird der letzte Anteil auf der linken Seite gleich Null, und die erhaltene Beziehung kann nach der zweiten Ableitung

$$\partial^2\lambda / \partial p_k \, \partial p_j$$

umgestellt werden. Durch Multiplikation mit Δp_k und Δp_j und Summation ergibt sich

$$\begin{aligned} & V^{0*} B^0 U^0 \sum_{k,j=1}^{s} \Delta p_k \Delta p_j \frac{\partial^2 \lambda}{\partial p_k \, \partial p_j} \\ & = 2V^{0*}((A(p) - \lambda^0 B(p)) \Delta U - 2\,\Delta\lambda\, V^{0*}(B(p) - B^0) U^0 - 2\,\Delta\lambda\, V^{0*} B^0 \Delta U . \end{aligned} \tag{14}$$

Aus dieser Formel kann

$$\Delta_2 \lambda = \sum_{k,j=1}^{s} \Delta p_k \Delta p_j \frac{\partial^2 \lambda}{\partial p_k \, \partial p_j}$$

ermittelt werden, wobei die oben hergeleiteten Korrekturterme $\Delta\lambda$ und ΔU Verwendung finden. In Ergänzung zu Formel (12) kann dann

$$\lambda(p) \approx \lambda(p^0) + \Delta\lambda + \tfrac{1}{2}\Delta_2\lambda \tag{15}$$

gesetzt werden.

Die Formel (14) ist nur für Matrizen A und B mit linearer Abhängigkeit von den Parametern gemäß (7) gültig, bei nichtlinearer Abhängigkeit dieser Matrizen vom Parametervektor p entstehen kompliziertere Formeln, in die analog zu Formel (10) Ableitungen der Matrizen A und B nach den Parametern eingehen.

3. Anwendungen auf gedämpfte und ungedämpfte Systeme

Betrachtet man ungedämpfte Systeme, so sind die Matrizen A (Steifigkeitsmatrix) und B (Trägheitsmatrix) in der Aufgabenstellung (1) positiv definit und symmetrisch. In diesem Fall braucht man nicht zwischen Links- und Rechts-Eigenvektoren zu unterscheiden. Setzt man entsprechend (7) eine lineare Abhängigkeit der Matrizen A und B von den Parametern p_k voraus, so kann man in Übereinstimmung mit praktischen Gegebenheiten annehmen, daß A_0 und B_0 positiv definit, A_k und B_k, $k = 1, \ldots, s$, nichtnegativ und die Parameter p_k ebenfalls nichtnegativ sind. Für die Eigenwertkorrektur kann man dann anstelle (11) die einfachere Formel

$$\Delta\lambda = U^{0\mathrm{T}}(A(p) - \lambda^0 B(p))\, U^0$$

erhalten, denn für das Eigenvektorensystem wird gewöhnlich B^0-Orthonormiertheit vorausgesetzt, so daß $U^{0\mathrm{T}}B^0U^0 = 1$ zu setzten ist.

Auch für den quadratischen Korrekturanteil $\Delta_2\lambda$ in Formel (14) kann man eine weitere Vereinfachung erzielen, wenn man Formel (13) transponiert,

$$U^{0\mathrm{T}}(A(p) - \lambda^0 B(p)) = \Delta\lambda\, U^{0\mathrm{T}}B^0 - (\Delta U)^\mathrm{T}\,(A^0 - \lambda^0 B^0),$$

und in (14) einsetzt. Es ergibt sich

$$\Delta_2\lambda = -2(\Delta U)^\mathrm{T}\,(A^0 - \lambda^0 B^0)\,\Delta U - 2\,\Delta\lambda\, U^{0\mathrm{T}}(B(p) - B^0)\, U^0\,.$$

Der Vektor ΔU wird aus dem System (13) bestimmt.

In der Praxis zerfällt der Parametervektor p häufig in zwei Anteile $p = (c, m)^\mathrm{T}$, wobei $c = (c_1, \ldots, c_q)^\mathrm{T}$ die Parameter der Steifigkeitsmatrix und $m = (m_1, \ldots, m_t)^\mathrm{T}$ die Parameter der Trägheitsmatrix enthält.

Die Untersuchung von gedämpften Systemen erfordert eine Auswertung der Formeln (3)–(6). Die Matrizen $M(p)$, $D(p)$ und $C(p)$ seien vom Parametervektor $p = (p_1, \ldots, p_s)^\mathrm{T}$ abhängig, der in praktischen Anwendungsbeispielen häufig in drei Anteile $p = (c, d, m)^\mathrm{T}$ zerfällt (Parameter c_i in der Steifigkeitsmatrix, d_j in der Dämpfungsmatrix und m_k in der Trägheitsmatrix). Setzen wir analog zu (7) lineare Abhängigkeit voraus, so sind auch die nach Formel (5) gebildeten Matrizen A und B linear von den Parametern p_k abhängig.

Die Eigenwerte λ der Aufgabe (3) treten bekanntlich als Paare konjugiert komplexer Zahlen auf. Es seien X^0 und Y^0 die Rechts- bzw. Links-Eigenvektoren der Aufgabe (3) und (4), wobei λ^0 eine Lösung der Aufgabe

$$\det\,(\lambda^2 M(p^0) + \lambda D(p^0) + C(p^0)) = 0$$

ist. Dann sind X^0 und Y^0 nichttriviale Lösungen der homogenen Systeme $Z(\lambda^0, p^0)\,X = 0$ bzw. $Z^*(\lambda^0, p^0)\;Y = 0$, wobei $Z(\lambda, p) = \lambda^2 M(p) + \lambda D(p) + C(p)$ bezeichnet wurde.

Um den linearen Korrekturterm $\Delta\lambda$ für $\lambda(p)$ zu ermitteln, ergeben sich aus Formel (11) unter Verwendung der Formeln (5) und (6) folgende Ausdrücke:

$$V^{0*}(A(p) - \lambda^0 B(p))\, U^0 = -Y^{0*}(\lambda^{02} M(p) + \lambda^0 D(p) + C(p))\, X^0 = -Y^{0*}Z(\lambda^0, p)\, X^0\,,$$

$$V^{0*}B^0U^0 = \frac{1}{\lambda^0}\, Y^{0*}(\lambda^{02}M^0 - C^0)\, X^0\,.$$

Hier wurde $M^0 = M(p^0)$ und $C^0 = C(p^0)$ bezeichnet. Dann folgt

$$\Delta\lambda = -\lambda^0\,\frac{Y^{0*}Z(\lambda^0, p)\, X^0}{Y^{0*}(\lambda^{02}M^0 - C^0)\, X^0}\,. \tag{16}$$

Setzt man in Formel (13) $\Delta U = (\Delta U_1, \Delta U_2)^\mathrm{T}$, so erhält man mit der Bezeichnung $D^0 = D(p^0)$ das System

$$\begin{aligned} &-\lambda^0\,\Delta U_1 + \Delta U_2 = \Delta\lambda X^0\,,\\ &-C^0\,\Delta U_1 - \lambda^0 M^0\,\Delta U_2 - D^0\,\Delta U_2 = \Delta\lambda\,\lambda^0 M^0 X^0 + Z(\lambda^0, p)\, X^0\,. \end{aligned} \tag{17}$$

Stellt man die erste Gleichung nach ΔU_2 um und setzt in die zweite Gleichung ein, so ergibt sich ein System zur Bestimmung des Vektors ΔU_1 in der Form

$$Z(\lambda^0, p^0) \, \Delta U_1 = -\Delta\lambda(2\lambda^0 M^0 + D^0) \, X^0 - Z(\lambda^0, p) \, X^0 \tag{18}$$

mit singulärer Systemmatrix $Z(\lambda^0, p^0)$. Es ist unter Nutzung von Formel (16) unschwer zu sehen, daß die rechte Seite des Systems (18) orthogonal zu Y^0 liegt, so daß dieses System sinnvoll gestellt ist und unter ΔU_1 die normkleinste Lösung von (18) zu verstehen ist.

Damit ergibt sich für die lineare Korrektur des Eigenvektors $X(p)$ die Beziehung

$$X(p) \approx X^0 + \Delta U_1 .$$

Wir vermerken, daß der Anteil ΔU_2 des Vektors ΔU zunächst nicht benötigt wird. Verfolgen wir jedoch das Ziel, entsprechend Formel (14) und (15) auch den quadratischen Korrekturanteil $\frac{1}{2} \Delta_2 \lambda$ für den Eigenwert zu berechnen, so wird der gesamte Vektor benötigt, der mit Hilfe von ΔU_1 aus der ersten Gleichung des Systems (17) beschafft werden kann.

4. Inverse Aufgabenstellung

Die hier erzielten Resultate der Sensibilitätsberechnung für die Eigenwerte der Aufgabe (1) unter der Voraussetzung (7) der linearen Abhängigkeit der Matrizen $A(p)$ und $B(p)$ vom Parametervektor p beinhalten die Möglichkeit, für eine Variante des inversen Eigenwertproblems ein iteratives Lösungsverfahren zu formulieren. Die direkte Eigenwertaufgabe (1) sei für $p = p^m$ (der obere Index m sei Iterationszähler) gelöst, und es seien $\lambda_1^m, \ldots, \lambda_n^m$ die Eigenwerte mit den dazugehörenden Systemen der Rechts- und Links-Eigenvektoren $U_1^m, \ldots, U_n^m$ bzw. $V_1^m, \ldots, V_n^m$.

Wir betrachten die folgende Variante für das inverse Eigenwertproblem (sie schließt andere Aufgabenformulierungen für die inverse Eigenwertaufgabe quadratischer Matrizen ein; siehe hierzu z. B. [1], dort auch weitere Literaturangaben). Für eine Indexteilmenge $\{k_1, \ldots, k_r\} \subset \{1, \ldots, n\}$ seien Eigenwertvorgaben formuliert, d. h. es seien Werte $\tilde{\lambda}_k$, $k = k_1, \ldots, k_r$, als Eigenwerte vorgegeben. Es sei nun die Aufgabe gestellt, den Parametervektor $p = (p_1, \ldots, p_s)^T$ so festzulegen, daß

$$\lambda_k(p) = \tilde{\lambda}_k , \qquad k = k_1, \ldots, k_r , \tag{19}$$

gilt (so können z. B. für ein ungedämpftes System die r kleinsten Eigenwerte vorgegeben sein).

Es ist zu bemerken, daß diese Aufgabe nur sinnvoll ist, wenn die Indizierung der Eigenwerte für alle in Betracht kommenden Parametervektoren p nach dem gleichen Prinzip vorgenommen wurde (für ungedämpfte Schwingungen mit parameterabhängigen Steifigkeits- und Trägheitsmatrizen kann man die Eigenwerte z. B. in wachsender Ordnung indizieren).

Weiterhin sei angemerkt, daß die Aufgabe (19) nicht immer lösbar ist bzw. sie braucht nicht eindeutig lösbar zu sein. Ersetzt man in Formel (19) die gesuchte Funktion $\lambda_k(p)$ durch ihren linearen Anteil gemäß (12), so ergibt sich anstelle (19) das System

$$\lambda_k(p^m) + \Delta^{(m)}\lambda_k = \tilde{\lambda}_k , \qquad k = k_1, \ldots, k_r ,$$

bzw.

$$\Delta^{(m)}\lambda_k = \tilde{\lambda}_k - \lambda_k(p^m) , \qquad k = k_1, \ldots, k_r . \tag{20}$$

In Verbindung mit der Voraussetzung der linearen Abhängigkeit der Matrizen $A(p)$ und $B(p)$ von p und Formel (11) ist (20) ein lineares System der s Variablen $p_1, \ldots, p_s$ mit einer i. a. nichtquadratischen Systemmatrix $S^{(m)}$. Bezeichnen wir mit $f^{(m)}$ die rechte Seite des Systems (20), so ist das System

$$S^{(m)} p = f^{(m)} \tag{21}$$

zu lösen, das in Abhängigkeit von s und r und von der konkreten Form der Systemdaten $S^{(m)}$ und $f^{(m)}$ eindeutig bzw. mehrdeutig lösbar, aber auch wiedersprüchlich sein kann. So wird i. a. anstelle des Systems (20) bzw. (21) eine Aufgabe der Art

$$\left.\begin{aligned} &\min_{p \in P_m} \|R^{(m)} p - g^{(m)}\|^2 , \\ &P_m = \left\{ p : \min_{u \in \mathbb{R}^s} \|S^{(m)} u - f^{(m)}\|^2 = \|S^{(m)} p - f^{(m)}\|^2 \right\} \end{aligned}\right\} \tag{22}$$

zu lösen sein. Hier bezeichnen $R^{(m)}$ und $g^{(m)}$ vorzugebenden Matrizen bzw. Vektoren (z. B. $R^{(m)} = I = \operatorname{diag}(1, \ldots, 1)$, $g^{(m)} = 0$ für die Ermittlung der normkleinsten Lösung des Systems (21)), die sich bei praktischen Problemen häufig aus dem physikalischen Zusammenhang ergeben und eine eindeutige Lösung der Aufgabe (22) festlegen. Wir definieren nun diese eindeutige Lösung der Aufgabe (22) als $(m + 1)$-te Iterierte p^{m+1}. Das beschriebene iterative Verfahren zur Realisierung der Eigenwertvorgaben (19) besteht aus einer Folge von Korrekturen des Parametervektors p, wobei nach jeder Korrektur dieses Vektors eine direkte Eigenwertaufgabe zu lösen ist (ähnliche Lösungsansätze siehe [4], [5], [6], [8]).

Wendet man dieses hier vorgeschlagene iterative Verfahren auf den Fall gedämpfter Schwingungen an, so wird die linke Seite des Systems (20) durch Formel (16) ersetzt, die linear in Bezug auf den Parametervektor p ist. Damit entstand ein iteratives Verfahren zur Lösung der inversen Aufgabe (19) für das quadratische Eigenwertproblem (3).

Es ist naheliegend, zur Lösung der inversen Aufgabe (19) nicht nur lineare Approximationen der Funktionen $\lambda_k(p)$ zu verwenden, sondern auf die in Formel (15) gegebenen quadratischen Approximationen zurückzugreifen. Das würde im Vergleich zu (21) auf ein System nichtlinearer Gleichungen führen, wodurch der numerische Aufwand sprungartig ansteigen würde (wenn auch für über- bzw. unterbestimmte nichtlineare Systeme handhabbare und stabile numerische Verfahren vorhanden sind [3]). Wie das nachfolgende Beispiel zeigt, ist die Verwendung von Formel (15) zur Lösung der inversen Aufgabe auch sachlich nicht immer zu rechtfertigen.

Es sei

$$A(p) = \begin{bmatrix} 1 & 0 & 0 \\ 0 & 3 & 0 \\ 0 & 0 & 8 \end{bmatrix} + p_1 \begin{bmatrix} 1 & 1 & 0 \\ 1 & 1 & 0 \\ 0 & 0 & 0 \end{bmatrix} + p_2 \begin{bmatrix} 0 & 0 & 0 \\ 0 & 1 & 1 \\ 0 & 1 & 1 \end{bmatrix},$$

$$B(p) = I = \operatorname{diag}(1, \dots, 1), \qquad p_1, p_2 \text{ reelle Parameter.}$$

Sei $p^0 = (0, 0)^{\mathrm{T}}$, dann besitzt die Aufgabe $(A(p^0) - \lambda B(p^0))\, U = 0$ die Eigenwerte $\lambda_1^0 = 1$, $\lambda_2^0 = 3$, $\lambda_3^0 = 8$. Es ergeben sich folgende Korrekturterme:

$$\Delta\lambda_1 = p_1, \qquad \tfrac{1}{2}\Delta_2\lambda_1 = -\tfrac{1}{2} p_1^2,$$

$$\Delta\lambda_2 = p_1 + p_2, \qquad \tfrac{1}{2}\Delta_2\lambda_2 = \tfrac{1}{2} p_1^2 - \tfrac{1}{5} p_2^2,$$

$$\Delta\lambda_3 = p_2, \qquad \tfrac{1}{2}\Delta_2\lambda_3 = \tfrac{1}{5} p_2^2.$$

Es seien die folgenden Eigenwertvorgaben für die zwei kleinsten Eigenwerte gegeben:

$$\tilde{\lambda}_1 = 2, \qquad \tilde{\lambda}_2 = 5.$$

Bei Verwendung linearer Approximationen für $\lambda_1(p)$ und $\lambda_2(p)$ erhalten wir gemäß (20) das System

$$p_1 = \tilde{\lambda}_1 - \lambda_1^0, \qquad p_1 + p_2 = \lambda_2 - \lambda_2^0,$$

aus dem sich die Lösung $p^1 = (p_1^1, p_2^1)^{\mathrm{T}} = (1, 1)^{\mathrm{T}}$ ergibt. Die Eigenwerte für die Aufgabe $(A(p^1) - \lambda B(p^1))\, U = 0$ sind $\lambda_1^1 = 1{,}6853$, $\lambda_2^1 = 5{,}0711$, $\lambda_3^1 = 9{,}2436$. Verwendet man dagegen die quadratischen Korrekturen (15), so ergibt sich das nichtlineare System

$$p_1 - \tfrac{1}{2} p_1^2 = \tilde{\lambda}_1 - \lambda_1^0 = 1, \qquad p_1 + p_2 + \tfrac{1}{2} p_1^2 - \tfrac{1}{5} p_2^2 = \tilde{\lambda}_2 - \lambda_2^0 = 2,$$

das offenbar keine reelle Lösung für p_1 besitzt.

5. Beispiel

Wir betrachten das Modell einer ungedämpften Schwingung mit der „Steifigkeitsmatrix"

$$C(c) = \begin{bmatrix} 2&0&0&0 \\ 0&3&0&0 \\ 0&0&5&0 \\ 0&0&0&8 \end{bmatrix} + c_1 \begin{bmatrix} 1&1&0&0 \\ 1&1&0&0 \\ 0&0&0&0 \\ 0&0&0&0 \end{bmatrix} + c_2 \begin{bmatrix} 0&0&0&0 \\ 0&1&1&0 \\ 0&1&1&0 \\ 0&0&0&0 \end{bmatrix} + c_3 \begin{bmatrix} 0&0&0&0 \\ 0&0&0&0 \\ 0&0&1&1 \\ 0&0&1&1 \end{bmatrix}$$

und der „Trägheitsmatrix"

$$M(m) = \begin{bmatrix} 1&0&0&0 \\ 0&1&0&0 \\ 0&0&1&0 \\ 0&0&0&1 \end{bmatrix} + m_1 \begin{bmatrix} 1&0&1&0 \\ 0&0&0&0 \\ 1&0&1&0 \\ 0&0&0&0 \end{bmatrix} + m_2 \begin{bmatrix} 0&0&0&0 \\ 0&1&0&1 \\ 0&0&0&0 \\ 0&1&0&1 \end{bmatrix}.$$

Wir setzen $c_k \geqq 0$ und $m_j \geqq 0$ voraus.

Der Parametervektor $p = (c, m)^{\mathrm{T}}$ zerfällt in die Anteile $c = (c_1, c_2, c_3)^{\mathrm{T}}$ und $m = (m_1, m_2)^{\mathrm{T}}$. Es sei $p^0 = (0, \dots, 0)^{\mathrm{T}}$. Mit Hilfe der Formel (11) ermitteln wir die linearen Korrekturanteile für die Eigenwerte:

$$\begin{aligned} \Delta\lambda_1 &= c_1 - \lambda_1 m_1, & \Delta\lambda_2 &= c_1 + c_2 - \lambda_2 m_2, \\ \Delta\lambda_3 &= c_2 + c_3 - \lambda_3 m_1, & \Delta\lambda_4 &= c_3 - \lambda_4 m_2. \end{aligned} \tag{23}$$

Man sieht hier deutlich, daß die Parameter m_j mit den Eigenwerten multipliziert werden, so daß ihr Einfluß bei großen Eigenwerten dominiert. In den quadratischen Korrekturtermen wird dieser starke Einfluß der großen Eigenwerte etwas gemildert, wie z. B. die Formel

$$\tfrac{1}{2}\Delta_2\lambda_4 = -(-\tfrac{1}{5}\lambda_4^2 m_2^2 - \tfrac{1}{3}c_3^2) - (c_3 - \lambda_4 m_2)\, m_2$$

zeigt, denn es ist $\frac{1}{5} = (\lambda_4 - \lambda_3)^{-1}$.

Es sei $c_1 = 0{,}10$, $c_2 = 0{,}15$, $c_3 = 0{,}05$, $m_1 = 0{,}10$, $m_2 = 0{,}05$. Dann ergibt sich die Tabelle 1.

Tabelle 1

λ_k	$\lambda_k + \Delta\lambda_k$	$\lambda_k + \Delta\lambda_k + \frac{1}{2}\Delta_2\lambda_k$	$\lambda_k(c, m)$
2	1,90	1,8840	1,8901
3	3,10	3,0730	3,0901
5	4,70	4,8237	4,8009
8	7,65	7,7003	7,6971

Wie die Formeln (23) zeigen, besitzt das System $\Delta\lambda_k = 0$ für mehrere Teilmengen $k \in \{k_1, \ldots\} \subset \{1, \ldots, 4\}$ nichtnegative Lösungen, so daß die entsprechenden Eigenwerte λ_k „relativ" stabilisiert werden können. Es können aber nicht alle vier Eigenwerte gleichzeitig stabilisiert werden, denn das System $\Delta\lambda_k = 0$, $k = 1, \ldots, 4$, besitzt außer der trivialen keine weitere nichtnegative Lösung. Wir vermerken abschließend, daß Eigenwertgruppen auch „quadratisch stabilisiert" werden können, denn es besitzen auch Systeme der Form $\Delta\lambda_k + \frac{1}{2}\Delta_2\lambda_k$, $k = k_1, \ldots$, nichtnegative Lösungen.

Literatur

1 FRIEDLAND, S.; NOCEDAL, J.; OVERTON, M. L.: The formulation and analysis of numerical methods for inverse eigenvalue problems. SIAM J. Numer. Anal. **24** (1987), 634–667.

2 DIEWALD, W.; KLIMMEK, A.; NORDMANN, R.: Die Berechnung des Biegeschwingungsverhaltens rotierender Maschinen. VDI-Berichte **786** (1989), 115–135.

3 HOFMANN, B.: Regularization for applied inverse and ill-posed problems. Teubner-Texte zur Mathematik, Bd. 85, Teubner, Leipzig 1986.

4 KUHNERT, F.: Über numerische Verfahren für inverse Matrizeneigenwertprobleme. Beiträge zur Numer. Math. **8** (1980), 75–84.

5 LUSCHNAT, E.: Modellierung eines inversen Eigenwertproblems durch eine nichtlineare Optimierungsaufgabe. Dissertation, TU Karl-Marx-Stadt 1989.

6 SCHWIBINGER, P.; NORDMANN, R.: Improvement of a reduced torsional model by means of parameter identification. J. Vibration, Stress and Reliability in Design **111** (1989), 17–26.

7 SCHWIBINGER, P.; NORDMANN, R.: The influence of torsional-lateral coupling on the stability behavior of geared rotor systems. J. Engineering for Gas Turbines and Power **110** (1988), 563–571.

8 STADLER, E.: Zur Lösung eines inversen Matrizeneigenwertproblems mit dem Regularisierungsprinzip und eine Anwendung in der Maschinendynamik. Dissertation, TH Karl-Marx-Stadt 1985.

Eingegangen: 21. Mai 1990

Anschrift: Prof. Dr. rer. nat. habil. FRIEDER KUHNERT, Technische Universität Chemnitz, Sektion Mathematik, PSF 964, Chemnitz, O-9010 Deutschland

BOOK REVIEWS

Stillwell, J., Mathematics and Its History. Berlin etc., Springer-Verlag 1989. X, 371 pp., 163 figs., DM 98,00. ISBN 3-540-96981-0 (Undergraduate Texts in Mathematics)

The reader of STILLWELL's book should note that the history of mathematics and mathematics and its history are not the same thing. It is not the author's purpose to present a concise history of mathematics but to focus on a unified view of undergraduate mathematics by means of its history. In this manner, history becomes a means of unifying and motivating mathematics by tracing a few dominant themes (of algebra, analytic, projectiv, differential, and non-euclidean geometry, calculus, mechanics, number theory, group theory, topology) from ancient times to the present day. STILLWELL explains his aim in Chapter 20 (p. 313), which gives a view of the twentieth century: "We have to admit that our conception of classical mathematics becomes clearer when presented in modern terms, and this is the best way to make it accessible to mathematicians who are not professional historians. As it will be evident by now, this is the point of view I have adopted in this book."

In general, classical mathematics is presented in modern terms and concepts. The historian can follow this procedure of carrying the present back into the past up to a certain point, the mathematician is enabled to trace and evaluate the historical roots of a problem he is working on quickly and easily. This gain from the mathematical point of view is won at the cost of a possible gap to grasp the differences between past and present if the modern language is rich enough to bring out the conceptual differences.

What I have in mind can be illustrated by the differences between classical calculus and that of modern non-standard analysis: In detail, if a Dedekind cut is an explanation of the continuous in terms of the discrete (thus p. 40), (a contradiction in itself), it follows that any attempt to justify calculus from this starting point must obscure the historical roots of the seventeenth century (existence of infinitesimals) and also the modern foundations of Laugwitz and Robinson (see p. 101).

Nevertheless, Stillwell's concept remains a matter of interest for both the mathematician and the historian.

Leipzig R. THIELE

Dym, H.; Goldberg, S.; Kaashoek, M.A.; Lancaster, P. (eds.), The Gohberg Anniversary Collection. Vol. I: The Calgary Conference and Matrix Theory Papers. Vol. II: Topics in Analysis and Operator Theory. Basel etc., Birkhäuser Verlag 1989. VII, 494 pp.; IX, 547 pp., 2-volume set sfr 228.–. ISBN 3-7643-2283-7 (OT 40/41)

Die Konferenzbände enthalten 45 Beiträge zur Operatortheorie, zur Theorie der Matrizen und zu deren Anwendungen. Jüngste Ergebnisse der Forschung zu einer breiten Palette von Themen aus diesen Gebieten werden vorgestellt, und die Liste der Autoren enthält die Namen vieler führender Spezialisten.

Band I enthält im ersten Teil das wissenschaftliche Programm der Konferenz, die aus Anlaß des 60. Geburtstages von I. GOHBERG stattfand, einen 40 Seiten umfassenden autobiographischen Bericht von I. GOHBERG, der aber auch über das Biographische hinaus von Interesse ist, und Erinnerungen von M. G. KREIN, von früheren Kollegen und Koautoren von I. GOHBERG.

Der zweite Teil von Band I enthält 16 Originalarbeiten zur Theorie der Matrizen und deren Anwendungen. Zur Theorie der Toeplitzmatrizen und anderer Klassen strukturierter Matrizen werden unter anderem Eigenschaften der Inversen und Probleme der effizienten numerischen Invertierung untersucht. Er werden Vervollständigungsprobleme für partiell definierte Matrizen, Fragen zur Stabilität und zur Klassifikation invarianter Unterräume, Probleme der Faktorisierung von Matrixpolynomen, Konvergenzeigenschaften von Rayleighquotient-Algorithmen für symmetrische Matrizenscharen und andere Probleme behandelt. Eine Vielzahl der vorgestellten Resultate findet Anwendung in der Systemtheorie.

Band II enthält 29 Originalarbeiten zur Theorie linearer Operatoren (Theorie der Kontraktionen, Interpolationsprobleme, charakteristische Funktionen von Operatoren, Wiener-Hopf-Determinanten, C^*-Algebren u. a.), zur Theorie nichtlinearer Operatoren (nichtlineare Toeplitzoperatoren, nichtlineare Lifting-Theoreme, Faktorisierung nichtlinearer Systeme), zu Differential- und singulären Integralgleichungen (Näherungsmethoden zur Lösung, Existenz von Lösungen kinetischer Gleichungen, singuläre Störungen u. a.), zur Geometrie von Banachräumen, zur komplexen Funktionstheorie und zur mathematischen Physik.

Berlin P. JONAS

Boyd, J. P., Chebyshev & Fourier Spectral Methods. Berlin etc., Springer-Verlag 1989. XVI, 798 pp., 82 figs., DM 168.–. ISBN 3-540-51487-2 (Lecture Notes in Engineering 49)

In dem vorliegenden Buch werden Spektralmethoden zur Lösung von Randwert-, Eigenwert- und instationären Problemen aus der Sicht des Anwenders dargestellt. Als Ansatzfunktionen der Galerkin- und Kollokationsverfahren werden trigonometrische Funktionen, Čebyšev-Polynome, Hermite-, Laguerre-, rationale Čebyšev-, sinc- und sphärische harmonische Funktionen benutzt.

Das Buch ging aus Vorlesungen des Autors an der Universität Michigan (USA) für Studenten höherer Studienjahre der verschiedensten Disziplinen, wie Raumfahrttechnik, Meteorologie, Meereskunde, angewandte Mechanik, Schiffbau und Kerntechnik, hervor, ist aber nicht auf ein spezielles Anwendungsgebiet ausgerichtet. Vielmehr werden Lösungsmethoden für typische Differentialgleichungsprobleme vorgestellt. Der Stil des Autors ist dem angesprochenen Leserkreis sehr angepaßt. In den Mittelpunkt der Betrachtungen werden Lösungsalgorithmen, tragende Ideen und praktische Erfahrungen gestellt. Auf mathematische Strenge und Kürze wird zugunsten einer leichteren Verständlichkeit verzichtet. Bei Beweisen wird oft auf die Literatur verwiesen.

Vom Leser werden Grundkenntnisse in numerischer Mathematik, Fourieranalysis und komplexer Funktionentheorie erwartet. Am Beginn des Buches werden grundlegende Tatsachen der Theorie der Fourierreihen, der Interpolation und der numerischen Integration dargestellt. Das Buch ist kein mathematisches Lehrbuch und keine Faktensammlung, sondern ein wertvoller Erfahrungsbericht eines Spezialisten auf dem Gebiet der Spektralmethoden. Dem Leser werden viele praktische Hinweise und Faustregeln vermittelt. Dieses Buch will ein „Reiseführer durch die Čebyšev-Stadt sein, wo die einzelnen Algorithmen und Identitäten eine Gemeinschaft bilden. Hier sind die speziellen Funktionen spezielle Freunde. Dabei hat eine Differentialgleichung stets eine Pseudospektralmatrix in Begleitung. Die Programmstruktur Gitterpunkte – Basisfunktionen – Kollokationsmatrix ist so grundlegend für das Leben wie Wolke – Regen – Fluß – Meer." Der Autor benutzt oft bildhafte Vergleiche, was den Reiz der Lektüre erhöht. Das Buch ist breit geschrieben, enthält zahlreiche Abbildungen und Beispiele. Der Autor ist immer bestrebt, mit minimalem Formelaufwand das Wesentliche verständlich zu machen. Dies erklärt den Umfang von fast 800 Seiten, unter denen man aber ein Symbol- und vor allem ein Sachwortverzeichnis vergeblich sucht.

Das vorliegende Buch kann jedem Physiker, Ingenieur und Studenten, der sich für Spektralmethoden und deren Anwendung auf Differentialgleichungsprobleme interessiert, empfohlen werden. Auch Mathematiker können diesem Buch viele Anregungen für die Forschung entnehmen.

Rostock H. TASCHE

ZAMM · Z. angew. Math. Mech. 71 (1991) 7/8, 241–250 Akademie Verlag

SCHNEIDER, K. R.; HEINZ, G.

Branch Inclusion in Generic Hopf Bifurcation: General Approach

Dedicated to the memory of LOTHAR COLLATZ

Unter Verwendung von Mittelungstransformationen werden Näherungen höherer Ordnung für die Familie der abzweigenden periodischen Lösungen im Fall der generischen Hopf-Bifurkation bestimmt. Diese Näherungen gestatten die Anwendung des Satzes von Newton-Kantorovič zur Ableitung von Einschließungen von Amplitude und Periode.

By using averaging transformations we determine approximations of higher order for the family of bifurcating periodic solutions in the generic case of Hopf bifurcation. These approximations permit the application of the Newton-Kantorovič Theorem to derive inclusions of the amplitude and period.

Для типичного случая бифуркации Хопфа, с помощью осредняющих отображений определяются аппроксимации высшего порядка для семейства разветляющих периодических решений. Эти аппроксимации позволяют применения теоремы канторовичаньютона для вывода включения амплитуды и периода.

MSC (1980): 58F14

1. Introduction

There is a lot of real life processes which can be described in an appropriate Banach space by an autonomous evolution equation

$$du/dt = F(u, \lambda) \tag{1.1}$$

depending on a parameter. We assume (1.1) to possess a family $\bar{u}(\lambda)$ of stationary solutions and investigate the exchange of their stability when λ crosses a critical value λ_0. Throughout the following we assume $\lambda_0 = 0$. In general, this exchange of stability is connected with the appearance of exactly one family $\{p(t, \lambda)\}$ of periodic solutions from the stationary solution $\bar{u}(0)$. This phenomenon is referred to as Hopf-bifurcation. There is a vast literature to this kind of bifurcation [2–5, 7–16].

Under some conditions the Hopf-bifurcation problem for an equation in an infinite-dimensional Banach space can be reduced to the corresponding one for a two-dimensional system (1.1). This can be done via center manifold theory or Lyapunov-Schmidt procedure [1, 5, 9, 14, 16].

If we suppose that the bifurcating periodic solution $p(t, \lambda)$ exists for $\lambda > 0$ then concerning the asymptotic behaviour of the bifurcating periodic solution $p(t, \lambda)$ with respect to λ it is well known that in the generic case the amplitude $\| p(., \lambda)\|$ and the minimal (primitive) period $T(\lambda)$ admit the asymptotic expansions

$$\| p(., \lambda)\| := \max_{0 \leqq t \leqq T(\lambda)} |p(t, \lambda)| = a_1 \sqrt{\lambda} + a_2\lambda + \dots, \qquad T(\lambda) = b_0 + b_1\lambda + \dots .$$

The aim of our paper is to derive inclusions for $\| p(., \lambda)\|$ and $T(\lambda)$ in the form

$$0 < \pi_2(\lambda) \leqq \| p(., \lambda)\| \leqq \pi_1(\lambda), \qquad \tau_2(\lambda) \leqq T(\lambda) \leqq \tau_1(\lambda)$$

for sufficiently small $|\lambda|$. Our way to establish these inclusions is based on the Newton-Kantorovič theorem and some average transformations. An application of this approach to van der Pol's equation can be found in [13].

2. Assumptions. Reduction to a finite dimensional problem

We consider the equation (1.1) in a real Banach space X under the following assumptions:

(A_1). $F \in C^{5;2}_{u;\lambda}(\Omega \times \Lambda_0, X)$ where Ω is a neighbourhood of the origin in X; Λ_j, $j = 0, 1, \dots$, is the interval defined by $\Lambda_j := \{\lambda \in \mathbb{R}: |\lambda| < \lambda_j, \lambda_j > 0\}$.

(A_2). $F(0, \lambda) \equiv 0$ for all $\lambda \in \Lambda_0$.

(A_3). The Fréchet derivative $B := F_u(0, 0)$ is a sectorial operator.

(A_4). B has a compact resolvent.

(A_5). $\sigma(B) \cap \{\text{Re } z = 0\} = \{\pm bi, b > 0\}$, where $\sigma(B)$ denotes the spectrum of B.

(A_6). $\pm bi$ are simple eigenvalues of B.

Under these assumptions the Hopf bifurcation problem for (1.1) may be reduced by means of a center manifold (see [1, 10]) to the Hopf bifurcation problem for a two-dimensional system of ordinary differential equations

$$\mathrm{d}x/\mathrm{d}t = f(x, \lambda) \tag{2.1}$$

where f satisfies

(V_1). $f \in C^{5,2}_{x,\lambda}(G \times \Lambda_0, \mathbb{R}^2)$ where G is a neighbourhood of the origin in $\mathbb{R}^2$.

(V_2). $f(0, \lambda) \equiv 0$ for all $\lambda \in \Lambda_0$.

(V_3). There is a positive number $\lambda_1, \lambda_1 \leqq \lambda_0$, such that for $\lambda \in \Lambda_1$ the Jacobian $f_x(0, \lambda)$ has complex eigenvalues $e_{1,2}(\lambda) := \alpha(\lambda) \pm i\beta(\lambda)$ satisfying

$$\alpha(0) = 0\,, \qquad \beta(0) = b > 0\,. \tag{2.2}$$

Let $\alpha_3(\lambda)$ be the first focus number of the equilibrium point $x = 0$ of (2.1) (for definition see [11]). We assume the generic condition

(V_4): $\alpha'(0)\,\alpha_3(0) \neq 0$.

For definiteness we suppose

$$\alpha'(0)\,\alpha_3(0) < 0\,. \tag{2.3}$$

This condition implies that the bifurcating family of periodic solutions $\{p(t, \lambda)\}$ exists for $\lambda > 0$.

From (V_1)–(V_3) it follows that the components of f may be represented in the form

$$f_1(x, \lambda) := \alpha(\lambda)\,x_1 - \beta(\lambda)\,x_2 + g_1(x, \lambda)\,, \qquad f_2(x, \lambda) := \beta(\lambda)\,x_1 + \alpha(\lambda)\,x_2 + g_2(x, \lambda) \tag{2.4}$$

where for $(x, \lambda) \in G \times \Lambda_1$

$$g_1(x, \lambda) := \sum_{2 \leqq i+j \leqq 4} a^{(1)}_{ij}(\lambda)\,x_1^i x_2^j + h_1(x, \lambda)\,, \qquad g_2(x, \lambda) := \sum_{2 \leqq i+j \leqq 4} a^{(2)}_{ij}(\lambda)\,x_1^i x_2^j + h_2(x, \lambda)\,, \tag{2.5}$$

$h_k(x, \lambda) = O(|x|^5)$ as $x \to 0$ uniformly in $\lambda \in \Lambda_1$ for $k = 1, 2$.

Introducing polar coordinates r, φ near $x = 0$ by $x_1 = r\cos\varphi$, $x_2 = r\sin\varphi$ we get from (2.4), (2.5) that the system (2.1) is equivalent to the system

$$\frac{\mathrm{d}r}{\mathrm{d}t} = \alpha(\lambda)\,r + \tilde{g}_1(r, \varphi, \lambda)\cos\varphi + \tilde{g}_2(r, \varphi, \lambda)\sin\varphi\,,$$

$$\frac{\mathrm{d}\varphi}{\mathrm{d}t} = \begin{cases} \beta(\lambda) + r^{-1}[\tilde{g}_1(r, \varphi, \lambda)\cos\varphi + \tilde{g}_2(r, \varphi, \lambda)\sin\varphi] & \text{for} \quad r \neq 0 \\ \beta(\lambda) & \text{for} \quad r = 0 \end{cases} \tag{2.6}$$

where $\tilde{g}_k(r, \varphi, \lambda) := g_k(r\sin\varphi, r\cos\varphi, \lambda)$, $k = 1, 2$.

According to (V_2), (V_3) it follows from (2.6) that there are numbers r_1 and λ_2, $0 < \lambda_2 < \lambda_1$, such that for $\lambda \in \Lambda_2$, $0 \leqq r \leqq r_1$ the system (2.1) is equivalent to the differential equation

$$\mathrm{d}r/\mathrm{d}\varphi = k(r, \varphi, \lambda)\,, \tag{2.7}$$

where k is 2π-periodic in φ and has the same regularity properties as f. Thus k may be represented in the form

$$k(r, \varphi, \lambda) = \sum_{i=1}^{4} k_i(\varphi, \lambda)\,r^i + r^4\bar{k}(r, \varphi, \lambda) \tag{2.8}$$

where $k_i \in C^{52}_{\varphi\lambda}(R \times \Lambda_2, \mathbb{R})$ for $1 \leqq i \leqq 4$, $r^4\bar{k} \in C^{552}_{r\varphi\lambda}((0, r_1) \times \mathbb{R} \times \Lambda_2, \mathbb{R})$, $\bar{k}(r, \varphi, \lambda) = o(1)$ as $r \to 0$ uniformly in $\varphi \in \mathbb{R}$, $\lambda \in \Lambda_2$.

Furthermore, we have

$$k_i(\varphi, \lambda) = k_i(\varphi, 0) + k'_{i,\lambda}(\varphi, 0)\,\lambda + \lambda^2\tilde{k}_i(\varphi, \lambda)\,, \tag{2.9}$$

where for $1 \leqq i \leqq 4$

$$k_i(\varphi, 0),\ k'_{i,\lambda}(\varphi, 0) \in C^5(\mathbb{R}, \mathbb{R})\,, \qquad \lambda^2\tilde{k}_i(\varphi, \lambda) \in C^{52}_{\varphi\lambda}(\mathbb{R} \times \Lambda_2, \mathbb{R})\,,$$

$$|\tilde{k}_i(\varphi, \lambda)| \leqq m_i \quad \text{for all} \quad (\varphi, \lambda) \in \mathbb{R} \times \Lambda_2 \quad \text{for some } m_i > 0.$$

For abbreviation we set

$$b_i(\varphi) := k_i(\varphi, 0)\,, \qquad c_i(\varphi) := k'_{i,\lambda}(\varphi, 0)\,, \qquad 1 \leqq i \leqq 4\,. \tag{2.10}$$

A straightforward computation yields

$$b_1(\varphi) \equiv 0, \qquad c_1(\varphi) := \alpha'(0)/\beta(0), \tag{2.11}$$

$$b_2(\varphi) = (\beta(0))^{-1} [a_{20}^{(1)}(0) \cos^3 \varphi + (a_{11}^{(1)}(0) + a_{20}^{(2)}(0)) \cos^2 \varphi \sin \varphi + (a_{02}^{(1)}(0) + a_{11}^{(2)}(0)) \cos \varphi \sin^2 \varphi + a_{02}^{(2)}(0) \sin^3 \varphi]. \tag{2.12}$$

For $b_3(\varphi)$ see [12].

In the frame of Hopf bifurcation theory conditions are well-known guaranteeing the existence of a family $r_p(\varphi, \lambda)$ of 2π-periodic solutions of (2.7) such that $r_p(\varphi, \lambda) \to 0$ as $\lambda \to 0$. Our purpose is to construct an approximation $\bar{r}_p(\varphi, \lambda)$ for $r_p(\varphi, \lambda)$ and to estimate the error of $\bar{r}_p(\varphi, \lambda)$. The aim of the following transformations is to bring our original problem in a form allowing the application of the Newton-Kantorovič-Theorem in order to get an error estimate for the approximate solution.

3. Transformation of the basic equation

First we scale the variable r by

$$r = \varepsilon \bar{r}, \qquad \varepsilon = \lambda^{1/2}, \qquad \lambda \geqq 0, \qquad \varepsilon \geqq 0. \tag{3.1}$$

Using (3.1) we get from (2.7)–(2.10)

$$\mathrm{d}\bar{r}/\mathrm{d}\varphi = \varepsilon b_2(\varphi) \bar{r}^2 + \varepsilon^2 \bar{r} \left(\frac{a'(0)}{\beta(0)} + b_3(\varphi) \bar{r}^2 \right) + \varepsilon^3 \bar{r}^2 (c_2(\varphi) + b_4(\varphi) \bar{r}^2) + \tilde{q}_1(\bar{r}, \varphi, \varepsilon) \tag{3.2}$$

where $\tilde{q}_1(\bar{r}, \varphi\, \varepsilon) = O(\varepsilon^4)$ as $\varepsilon \to 0$ uniformly in $\varphi \in \mathbb{R}$, $0 \leqq \bar{r} \leqq \bar{r}^*$ for any fixed $\bar{r}^* > 0$.

By this way, the problem of bifurcation of a family of 2π-periodic solutions of the equation (2.7) from $r = 0$ is reduced to the problem of bifurcation of a family of 2π-periodic solutions of equation (3.1) from a circle $r = \text{const}$. In what follows, by means of averaging transformations we want to achieve a differential equation of the form

$$\mathrm{d}v/\mathrm{d}\varphi = \varepsilon^m (\beta_0(v) + \varepsilon \beta_1(v) + \bar{\beta}(v, \varphi, \varepsilon)) \tag{3.3}$$

where $\beta_0(v) \neq 0$, m is a natural number and $\bar{\beta}(v, \varphi, \varepsilon) = O(\varepsilon^2)$ as $\varepsilon \to 0$ uniformly in $\varphi \in \mathbb{R}$, $|v| \leqq v^*$ for any fixed v^*.

To this end we introduce the averaging operator M with respect to a continuous 2π-periodic function h by

$$Mh := \frac{1}{2\pi} \int_0^{2\pi} h(s)\, \mathrm{d}s.$$

It is obvious that we may represent any continuous 2π-periodic function h in the form

$$h(\varphi) = Mh + \frac{\mathrm{d}q}{\mathrm{d}\varphi}(\varphi) \tag{3.4}$$

where the 2π-periodic function q is defined by

$$q(\varphi) := \int_0^{\varphi} h(\sigma)\, \mathrm{d}\sigma - \varphi Mh.$$

At the first step we chose $h(\varphi) = b_2(\varphi)$. From (2.12) it follows $Mb_2 = 0$. Thus we have

$$b_2(\varphi) = \mathrm{d}q_1(\varphi)/\mathrm{d}\varphi$$

with

$$q_1(\varphi) = \int_0^{\varphi} b_2(s)\, \mathrm{d}s. \tag{3.5}$$

The corresponding coordinate transformation reads

$$\eta = \bar{r} - \varepsilon \bar{r}^2 q_1(\varphi) \tag{3.6}$$

which is one-to-one for $0 \leqq \bar{r} \leqq \bar{r}^*$, $0 \leqq \varepsilon \leqq \varepsilon_0(\bar{r}^*)$, where $\bar{r}^*$ is any fixed positive number. We write the corresponding inverse transformation in the form

$$\bar{r} = \eta + \varepsilon \eta^2 q_1(\varphi) + 2\varepsilon^2 \eta^3 (q_1(\varphi))^2 + \tilde{r}(\eta, \varphi, \varepsilon). \tag{3.7}$$

Using the Newton-Kantorovič-Theorem (see Proposition A in the Appendix) we get

$$\tilde{r}(\eta, \varphi, \varepsilon) = O(\varepsilon^3) \quad \text{as} \quad \varepsilon \to 0 \tag{3.8}$$

uniformly in $\varphi \in \mathbb{R}$, $0 \leqq \eta \leqq \eta^*$ for any fixed positive η^*.

From (3.2), (3.6)–(3.8) we get

$$\frac{d\eta}{d\varphi} = \varepsilon^2 \left\{ \frac{\alpha'(0)}{\beta(0)} \eta + (b_3(\varphi) - 2b_2(\varphi)\, q_1(\varphi))\, \eta^3 \right\} + \varepsilon^3 \left\{ \left(c_2(\varphi) - \frac{\alpha'(0)}{\beta(0)} q_1(\varphi) \right) \eta^2 + (b_4(\varphi) + b_3(\varphi)\, q_1(\varphi) - 6b_2(\varphi)\, (q_1(\varphi))^2)\, \eta^4 \right\} + \tilde{q}_2(\eta, \varphi, \lambda) \tag{3.9}$$

where

$$\tilde{q}_2(\eta, \varphi, \varepsilon) = O(\varepsilon^3) \quad \text{as} \quad \varepsilon \to 0 \tag{3.10}$$

uniformly in $\varphi \in \mathbb{R}$, $|\eta| \leqq \eta^*$.

At the second step we choose

$$h(\varphi) := b_3(\varphi) - 2b_2(\varphi)\, q_1(\varphi)\,.$$

The corresponding function $q := q_2$ in the representation (3.4) reads

$$q_2(\varphi) := \int_0^{\varphi} b_3(s)\, ds - \varphi M b_3 - (q_1(\varphi))^2\,.$$

Now we apply the coordinate transformation

$$\sigma = \eta - \varepsilon^2 q_2(\varphi)\, \eta^3 \tag{3.11}$$

which is one-to-one for $0 \leqq \eta \leqq \eta^*$, $0 \leqq \varepsilon \leqq \varepsilon_1(\eta^*) \leqq \varepsilon_0(\tilde{r}^*)$, where η^* is any fixed positive number. The corresponding inverse can be represented in the form

$$\eta = \sigma + \varepsilon^2 q_2(\varphi)\, \sigma^3 + \tilde{\eta}(\sigma, \varphi, \varepsilon) \tag{3.12}$$

where according to Proposition A

$$\tilde{\eta}(\sigma, \varphi, \varepsilon) = O(\varepsilon^4) \quad \text{as} \quad \varepsilon \to 0 \tag{3.13}$$

uniformly in $\varphi \in \mathbb{R}$, $0 \leqq \sigma \leqq \sigma^*$ for any fixed positive σ^*. Thus, we obtain from (3.9)

$$\frac{d\sigma}{d\varphi} = \varepsilon^2 \left\{ \frac{\alpha'(0)}{\beta(0)} + \sigma^2 M b_3 \right\} \sigma + \varepsilon^3 \left\{ \left(c_2(\varphi) - \frac{\alpha'(0)}{\beta(0)} q_1(\varphi) \right) \sigma^2 + (b_4(\varphi) + b_3(\varphi)\, q_1(\varphi) - 6b_2(\varphi)\, (q_1(\varphi))^2)\, \sigma^4 \right\} + \tilde{q}_3(\sigma, \varphi, \varepsilon) \tag{3.14}$$

with

$$\tilde{q}_3(\sigma, \varphi, \varepsilon) = O(\varepsilon^4) \quad \text{as} \quad \varepsilon \to 0 \tag{3.15}$$

uniformly in $\varphi \in \mathbb{R}$, $0 \leqq \sigma \leqq \sigma^*$ for any fixed σ^*.

To perform the last transformation we introduce the functions

$$h_3(\varphi) := c_2(\varphi) - \frac{\alpha'(0)}{\beta(0)} q_1(\varphi)\,, \tag{3.16}$$

$$h_4(\varphi) := b_4(\varphi) + b_3(\varphi)\, q_1(\varphi) - 6b_2(\varphi)\, (q_1(\varphi))^2\,, \tag{3.17}$$

$$q_3(\varphi) := \int_0^{\varphi} h_3(s)\, ds - \varphi M h_3\,, \qquad q_4(\varphi) := \int_0^{\varphi} h_4(s)\, ds - \varphi M h_4\,.$$

By means of the coordinate transformation

$$\nu = \sigma - \varepsilon^3 (q_3(\varphi) + q_4(\varphi)\, \sigma^2)\, \sigma^2 \tag{3.18}$$

which is one-to-one for $0 \leqq \sigma \leqq \sigma^*$, $0 \leqq \varepsilon \leqq \varepsilon_2(\sigma^*) \leqq \varepsilon_1(\eta^*)$, where σ^* is any fixed positive number. We represent the inverse in the form

$$\sigma = \nu + \varepsilon^3 \nu^3 q_3(\varphi) + \tilde{\sigma}(\nu, \varphi, \varepsilon)\,,$$

where

$$\tilde{\sigma}(\nu, \varphi, \varepsilon) = O(\varepsilon^6) \quad \text{as} \quad \varepsilon \to 0 \tag{3.19}$$

uniformly in $\varphi \in \mathbb{R}$, $0 \leqq \nu \leqq \nu^*$ for any fixed $\nu^* > 0$.

By this way we get from (3.14) the equation

$$\frac{dv}{d\varphi} = \varkappa(v, \varphi, \varepsilon) := \varepsilon^2 \left\{\frac{\alpha'(0)}{\beta(0)} + v^2 Mb_3\right\} v + \varepsilon^3\{Mh_3 + v^2 Mh_4\} v^2 + \varepsilon^4 v\tilde{\varkappa}(v, \varphi, \varepsilon) \tag{3.20}$$

with

$$\tilde{\varkappa}(v, \varphi, \varepsilon) = O(1) \quad \text{as} \quad \varepsilon \to 0 \tag{3.21}$$

uniformly in $\varphi \in \mathbb{R}$, $0 \leqq v \leqq v^*$ for an fixed $v^* > 0$.

Definition 3.1. A function $v \in C((0, \varepsilon^*), \mathbb{R}^+)$ satisfying $\varkappa(v(\varepsilon), \varphi, \varepsilon) = O(\varepsilon^{i+1})$ uniformly in φ is called a *quasistationary solution* of (3.20) of order i.

From (3.20) it follows that a solution of

$$\alpha'(0)/\beta(0) + v^2 Mb_3 = 0 \tag{3.22}$$

yields a quasistationary solution of (3.20) of second order.

Taking into account the relation $Mb_3 = \alpha_3(0)/2\pi \neq 0$ (see [12]) and the hypotheses (V_4), the equation (3.22) has a unique positive solution

$$v_1 = \sqrt{-\alpha'(0)\, 2\pi/\beta(0)\, \alpha_3(0)}\,. \tag{3.23}$$

We obtain a third order quasistationary solution of (3.20) by solving the equation

$$\frac{\alpha'(0)}{\beta(0)} + v^2 Mb_3 + \varepsilon\,(Mh_3 + v^2 Mh_4)\, v = 0 \tag{3.24}$$

up to order ε. It is easy to verify that

$$\bar{v}_2(\varepsilon) = v_1 + \varepsilon v_2 \tag{3.25}$$

is a third order quasistationary solution of (3.20) where v_2 is defined by

$$v_2 := -\frac{Mh_3 + v_1^2 Mh_4}{2Mb_3}. \tag{3.26}$$

In the next section we derive conditons ensuring the existence of a neighbourhood of a quasistationary solution of (3.20) containing a unique nontrivial 2π-periodic solution of (3.20).

4. Error estimates for the Hopf approximations

Our way to derive approximations for the bifurcating family $\{r_p(\varphi, \lambda)\}$ of 2π-periodic solutions of the equation (2.7) and to estimate their error is as follows. First we apply the Newton-Kantorovič-Theorem to construct neighbourhoods of the quasistationary solutions v_1 and $\bar{v}_2(\varepsilon)$ containing a unique 2π-periodic solution of (3.20). Then we use the one-to-one map between r and v to derive Hopf approximations and corresponding error estimates.

Let $C^k_{2\pi}(\mathbb{R})$ denote the space of k-times continuously differentiable 2π-periodic functions y mapping $\mathbb{R}$ into $\mathbb{R}$, let $Y := C^1_{2\pi}(\mathbb{R})$, $Z := C^0_{2\pi}(\mathbb{R})$. We equipe Z with the usual maximum norm $\|.\|$ such that Z is a Banach space, Y is an open subset of Z. We introduce the operator P by

$$P(y, \varepsilon)(\varphi) := \frac{dy}{d\varphi} - \varkappa(y, \varphi, \varepsilon) \tag{4.1}$$

where $\varkappa$ is defined in (3.20). Under the assumptions (V_1)–(V_4) we have $P := Y \times (0, \varepsilon^*) \to Z$ where ε^* is a sufficiently small positive number. Let $\bar{y}(\varepsilon) \in Y$ be an approximate solution of the equation $P(y, \varepsilon) = 0$ for $\varepsilon \in (0, \varepsilon^*)$. The operator $\partial P/\partial y(\bar{y}(\varepsilon), \varepsilon)$ is invertible iff the equation

$$\frac{dv}{d\varphi} - \varkappa_y(\bar{y}(\varepsilon)(\varphi), \varphi, \varepsilon)\, v = \psi \tag{4.2}$$

has a unique solution in Y for any ψ in Z. It is easy to verify that under our conditions the linear differential eqution (4.2) has to any $\psi \in Z$ a unique solution $\bar{v}(\varphi; v_0, \varepsilon, \psi)$ satisfying $\bar{v}(0; v_0, \varepsilon, \psi) = v_0$. To given $y(\varepsilon)$, $0 < \varepsilon < \varepsilon^*$, let ζ_ε be the map defined by

$$\zeta_\varepsilon(\varphi) := \exp\left(\int_0^\varphi \varkappa_y(\bar{y}(\varepsilon)(\sigma), \sigma, \varepsilon)\, d\sigma\right). \tag{4.3}$$

Using this function we may represent $\bar{v}(\varphi; v_0, \varepsilon, \psi)$ in the form

$$\bar{v}(\varphi; v_0, \varepsilon, \psi) = \zeta_\varepsilon(\varphi)\left(v_0 + \int_0^\varphi (\zeta_\varepsilon(\tau))^{-1}\,\psi(\tau)\,d\tau\right). \tag{4.4}$$

Assuming

$$\zeta_\varepsilon(2\pi) \neq 1 \quad \text{for} \quad 0 < \varepsilon < \varepsilon^* \tag{4.5}$$

we can put

$$v_0 := \frac{\zeta_\varepsilon(2\pi)}{1 - \zeta_\varepsilon(2\pi)} \int_0^{2\pi} (\zeta_\varepsilon(\tau))^{-1}\,\psi(\tau)\,d\tau\,. \tag{4.6}$$

It can easily be checked that $\bar{v}(\varphi; v_0, \varepsilon, \psi)$ with v_0 defined in (4.6) is a 2π-periodic solution of (4.2).

Let $\bar{y}(\varepsilon)$ be a quasistationary solution of (3.20) of any order i, $i \geqq 2$. Then from (3.20), (3.21) and (4.2) it follows

$$\zeta_\varepsilon(\varphi) = 1 - \varepsilon^2 \frac{2\alpha'(0)}{\beta(0)}\,\varphi + \varepsilon^3 \tilde{\zeta}_\varepsilon(\varphi) \tag{4.7}$$

with

$$\tilde{\zeta}_\varepsilon(\varphi) = O(1) \quad \text{as} \quad \varepsilon \to 0 \quad \text{uniformly in } \varphi. \tag{4.8}$$

Thus, there are a positive number ε_3, $\varepsilon_3 \leqq \varepsilon_2$, and a strictly increasing function k_2: $(0, \varepsilon_3) \to \mathbb{R}^+$ such that $\partial P/\partial y(\bar{y}(\varepsilon), \varepsilon)$ is invertible for $0 < \varepsilon < \varepsilon_3$ and we have

$$\begin{aligned} \Gamma_\varepsilon\psi(\varphi) &:= (\partial P/\partial y(\bar{y}(\varepsilon), \varepsilon))^{-1}\,\psi(\varphi) \\ &= \zeta_\varepsilon(\varphi)\left[\frac{\zeta_\varepsilon(2\pi)}{1 - \zeta_\varepsilon(2\pi)} \int_0^{2\pi} (\zeta_\varepsilon(\tau))^{-1}\,\psi(\tau)\,d\tau + \int_0^\varphi \zeta_\varepsilon(\tau)\,\psi(\tau)\,d\tau\right] \end{aligned}$$

and

$$\|\Gamma_\varepsilon\| \leqq k_2(\varepsilon) \quad \text{with} \quad k_2 = O(\varepsilon^{-2}) \quad \text{as} \quad \varepsilon \to 0\,. \tag{4.9}$$

Especially, (4.9) is valid for $\bar{y}(\varepsilon) = v_1$ and $\bar{y}(\varepsilon) = \bar{v}_2(\varepsilon)$.

From (3.20)–(3.21) it follows that there exist a positive number ε_4, $\varepsilon_4 \leqq \varepsilon_3$, and strictly increasing functions $k_1^{(1)}$, $k_1^{(2)}$: $[0, \varepsilon_4] \to \mathbb{R}^+$ such that

$$\left.\begin{aligned} &\|P(v_1, \varepsilon)\| \leqq k_1^{(1)}(\varepsilon)\,, && k_1^{(1)}(\varepsilon) = O(\varepsilon^3)\,, \\ &\|P(\bar{v}_2(\varepsilon), \varepsilon)\| \leqq k_1^{(2)}(\varepsilon)\,, && k_1^{(2)}(\varepsilon) = O(\varepsilon^4)\,, \\ &k_1^{(2)}(\varepsilon) \leqq k_1^{(1)}(\varepsilon)\,. \end{aligned}\right\} \tag{4.10}$$

Let

$$\delta_0(\varepsilon) = v_1/2\,. \tag{4.11}$$

By (3.20), (3.21) there are a positive number ε_5, $\varepsilon_5 \leqq \varepsilon_4$, and a strictly increasing function k_3: $[0, \varepsilon_5] \to \mathbb{R}^+$ such that

$$\|\partial^2 P/\partial y^2(y, \varepsilon)\| \leqq k_3(\varepsilon) \quad \text{for} \quad \|y - v_1\| < \delta_0(\varepsilon)\,, \qquad k_3(\varepsilon) = O(\varepsilon^2) \quad \text{as} \quad \varepsilon \to 0\,. \tag{4.12}$$

For $i = 1, 2$ we set

$$\delta_1^{(i)}(\varepsilon) := \frac{1 - \sqrt{1 - 2k_1^{(i)}(\varepsilon)\,k_2^2(\varepsilon)\,k_3(\varepsilon)}}{k_2(\varepsilon)\,k_3(\varepsilon)}\,, \qquad \delta_2^{(i)}(\varepsilon) := \frac{1 + \sqrt{1 - 2k_1^{(i)}(\varepsilon)\,k_2^2(\varepsilon)\,k_3(\varepsilon)}}{k_2(\varepsilon)\,k_3(\varepsilon)}\,. \tag{4.13}$$

From (4.9)–(4.13) it follows

$$\delta_1^{(i)}(\varepsilon) = O(\varepsilon^i)\,, \qquad \delta_2^{(i)}(\varepsilon) = O(\varepsilon^{-i})$$

such that there is a positive ε_6, ε_6, $\leqq \varepsilon_5$, with

$$2k_1^{(i)}(\varepsilon)\,k_2^2(\varepsilon)\,k_3(\varepsilon) < 1 \quad \text{for} \quad i = 1, 2\,, \tag{4.14}$$

$$\delta_1^{(2)}(\varepsilon) < \delta_1^{(1)}(\varepsilon) < \delta_0(\varepsilon) = \frac{v_1}{2} < \delta_2^{(1)}(\varepsilon) < \delta_2^{(2)}(\varepsilon)\,. \tag{4.15}$$

Applying Theorem A from the Appendix we have the following result:

Theorem 4.1: *Assume that* $(V_1)-(V_4)$ *hold. Then, for* $0 \leqq \varepsilon \leqq \varepsilon_6$, *the equation* (3.20) *has a family* $\{v_p(\varphi, \varepsilon)\}$ *of* 2π*-periodic solutions satisfying*

$$\|v_1 - v_p(\varphi, \varepsilon)\| := \max_{0 \leqq \varphi \leqq 2\pi} |v_1 - v_p(\varphi, \varepsilon)| \leqq \delta_1^{(1)}(\varepsilon) = O(\varepsilon), \tag{4.16}$$

$$\max_{0 \leqq \varphi \leqq 2\pi} |v_2(\varepsilon) - v_p(\varphi, \varepsilon)| \leqq \delta_1^{(2)}(\varepsilon) = O(\varepsilon^2).$$

Moreover, $v_p(\varphi, \varepsilon)$ *is a unique* 2π*-periodic solution of* (3.20) *in the set* $S_{v_1/2}$.

Theorem 4.1 provides estimations of the amplitude of the family of 2π-periodic solutions $v_p(\varphi, \varepsilon)$ to equation (3.20) which bifurcate from the circle $v = v_1$ when ε crosses $\varepsilon = 0$. To get estimates for the corresponding family $\bar{r}_p(\varphi, \varepsilon)$ belonging to (3.2) we take into consideration the coordinate transformations

$$\eta = \Phi_1(\bar{r}, \varphi, \varepsilon), \qquad \sigma = \Phi_2(\eta, \varphi, \varepsilon), \qquad v = \Phi_3(\sigma, \varphi, \varepsilon)$$

defined in (3.6), (3.12), (3.18) which yield a relation

$$v = \bar{r} + \varepsilon \Phi_4(\bar{r}, \varphi, \varepsilon). \tag{4.17}$$

To any positive $\bar{r}_2$ there is a number ε_7, $0 < \varepsilon_7 \leqq \varepsilon_6$, such that v is a strictly increasing function of $\bar{r}$ for $0 \leqq \varepsilon \leqq \varepsilon_7$, $0 \leqq \varphi \leqq 2\pi$, $0 \leqq \bar{r} \leqq \bar{r}_2$. Solving the equations

$$v_1 + \delta_1^{(i)}(\varepsilon) = \bar{r} + \varepsilon \Phi_4(\bar{r}, \varphi, \varepsilon), \qquad v_1 - \delta_1^{(i)}(\varepsilon) = \bar{r} + \varepsilon \Phi_4(\bar{r}, \varphi, \varepsilon), \tag{4.18}$$

$i = 1, 2$, we obtain the desired inclusion of the amplitude of the corresponding 2π-periodic solution $\bar{r}_p(\varphi, \varepsilon)$ of (3.2). To this end we take

$$\gamma_\pm = v_1 + v_1^2 q_1(\varphi)\, \varepsilon \pm \delta_1^{(1)}(\varepsilon)$$

in case $i = 1$ and

$$\gamma_\pm = v_1 + \varepsilon(v_2 + q_1(\varphi)\, v_1^2) + \varepsilon^2 v_1[v_1^2 q_2(\varphi) + q_1(\varphi)\,(v_2 + q_1(\varphi)\, v_1^2)] \pm \delta_1^{(2)}(\varepsilon)$$

in case $i = 2$ as approximations for the zeroes in (4.18). By applying Theorem A from the Appendix we get the existence of functions $\varrho_1^{(i)}: [0, \varepsilon_8] \to \mathbb{R}^+$, $0 < \varepsilon_8 \leqq \varepsilon_7$, with $\varrho_1^{(i)}(\varepsilon) = 0(\varepsilon^i)$, $i = 1, 2$, such that for $0 \leqq \varepsilon \leqq \varepsilon_8$

$$\|v_1 - \bar{r}_p(., \varepsilon)\| \leqq \varrho_1^{(1)}(\varepsilon), \qquad \|v_1 + \varepsilon(v_2 + q_1(\varphi)\, v_1^2) - \bar{r}_p(., \varepsilon)\| \leqq \varrho_1^{(2)}(\varepsilon). \tag{4.19}$$

Taking into account the scaling (3.1) we have the result:

Theorem 4.2: *Suppose the assumptions* $(V_1)-(V_4)$ *hold. Then there are a sufficiently small number* λ^* *and monotone increasing functions*

$$\bar{\varrho}_1^{(i)}: [0, \lambda^*) \to \mathbb{R}^+, \qquad \bar{\varrho}_1^{(i)}(\lambda) = O(\lambda^{(i+1)/2}) \tag{4.20}$$

as $\lambda \to 0$, $i = 1, 2$, *such that for* $0 < \lambda < \lambda^*$ *equation* (2.7) *has a unique family of* 2π*-periodic solutions* $r = r_p(\varphi, \lambda)$ *satisfying*

$$\|v_1 \sqrt{\lambda} - r_p(., \lambda)\| \leqq \bar{\varrho}_1^{(1)}(\lambda), \qquad \|v_1 \sqrt{\lambda} + \lambda(v_2 + q_1(\varphi)\, v_1^2) - r_p(., \lambda)\| \leqq \bar{\varrho}_1^{(2)}(\lambda). \tag{4.21}$$

Remark: The approximation $v_1 \sqrt{\lambda}$ to the Hopf oscillation $r_p(\varphi, \lambda)$ is well-known. We call it the *first approximation of the amplitude* $a(\lambda)$. The function $v_1 \sqrt{\lambda} + \lambda(v_2 + q_1(\varphi)\, v_1^2)$ is said to be the *second order approximation of the Hopf oscillation*. It can be easily used to derive a second order inclusion of the amplitude $a(\lambda)$. Let $q := \|q_1\|$. Then we get

$$v_1 \sqrt{\lambda} + \lambda(v_2 - q v_1^2) - \bar{\varrho}_1^{(2)}(\lambda) \leqq a(\lambda) \leqq v_1 \sqrt{\lambda} + \lambda(v_2 + q v_1^2) + \bar{\varrho}_1^{(2)}(\lambda)$$

Higher order approximations can be obtained by means of quasistationary solutions of (3.20) of higher order.

5. Inclusion of the primitive period

For $\lambda \in \Lambda_2$ and $0 \leqq r \leqq r_1$ system (2.1) is equivalent to system (2.6) and to equation (2.7). To the family $\{r_p(\varphi, \lambda)\}$ of 2π-periodic solutions of (2.7) does correspond the family $\{p(t, \lambda)\}$ of periodic solutions of (2.1) bifurcating from the equilibrium point $x = 0$ when λ crosses $\lambda = 0$. Let $T(\lambda)$ be the primitive period of $p(t, \lambda)$. To estimate $T(\lambda)$ we use the second equation in (2.6). By using the function

$$\gamma(r, \varphi, \lambda) := \begin{cases} -r^{-1}[\tilde{f}_2(r, \varphi, \lambda) \cos \varphi - \tilde{f}_1(r, \varphi, \lambda) \sin \varphi] & \text{for } r \neq 0 \\ 0 & \text{for } r = 0 \end{cases} \tag{5.1}$$

we rewrite the second equation of (2.6) in the form

$$d\varphi/dt = \beta(\lambda) + \gamma(r, \varphi, \lambda). \tag{5.2}$$

From this equation we obtain

$$T(\lambda) = \int_0^{2\pi} \frac{d\varphi}{\beta(\lambda) + \gamma(r_p(\varphi, \lambda), \varphi, \lambda)}. \tag{5.3}$$

In what follows we use the estimate (4.21) of the amplitude of $r_p(\varphi, \lambda)$ to derive an inclusion for the primitive period $T(\lambda)$. Under our assumptions we obtain from (5.1), (2.5), (2.6) for $\lambda \in \Lambda_2$

$$\gamma(r, \varphi, \lambda) = r l_1(\varphi, \lambda) + r^2 l_2(\varphi, \lambda) + r^3 l_3(r, \varphi, \lambda), \tag{5.4}$$

where

$$l_1(\varphi, \lambda) := a_{20}^{(2)}(\lambda) \cos^3 \varphi + (a_{11}^{(2)}(\lambda) - a_{20}^{(1)}(\lambda)) \cos^2 \varphi \sin \varphi$$
$$+ (a_{02}^{(2)}(\lambda) - a_{11}^{(1)}(\lambda)) \cos \varphi \sin^2 \varphi - a_{02}^{(1)}(\lambda) \sin^3 \varphi, \tag{5.5}$$

$$l_2(\varphi, \lambda) := a_{30}^{(2)}(\lambda) \cos^4 \varphi + (a_{21}^{(2)}(\lambda) - a_{30}^{(1)}(\lambda)) \cos^3 \varphi \sin \varphi + (a_{12}^{(2)}(\lambda) - a_{21}^{(1)}(\lambda)) \cos^2 \varphi \sin^2 \varphi$$
$$+ (a_{03}^{(2)}(\lambda) - a_{12}^{(1)}(\lambda)) \cos \varphi \sin^3 \varphi - a_{03}^{(1)}(\lambda) \sin^4 \varphi. \tag{5.6}$$

An important property of $l_1(\varphi, \lambda)$ is

$$\int_0^{2\pi} l_1(\varphi, \lambda) \, d\varphi = 0.$$

Moreover, there are positive constants $c_i, i = 1, \ldots, 6, \lambda_3, r_2, \lambda_3 \leqq \lambda_2, r_2 \leqq r_1$, such that for $0 \leqq r \leqq r_2, 0 \leqq \lambda \leqq \lambda_3$, $0 \leqq \varphi \leqq 2\pi$,

$$l_i(\varphi, \lambda) = l_i(\varphi, 0) + \lambda \tilde{l}_i(\varphi, \lambda), \qquad |\tilde{l}_i(\varphi, \lambda)| \leqq c_i, \qquad i = 1, 2. \tag{5.7}$$

$$|l_3(r, \varphi, \lambda)| \leqq c_3, \tag{5.8}$$

$$\beta^{-1}(\lambda) = \beta^{-1}(0) - \beta^{-1}(0) \beta'(0) \lambda + \lambda^2 \beta_r(\lambda), \qquad |\beta_r(\lambda)| \leqq c_4. \tag{5.9}$$

Hence, there are positive constants λ_4, r_3, $\lambda_3 \leqq \lambda_2$, $r_2 \leqq r_1$, such that

$$|\gamma(r, \varphi \lambda)/\beta(\lambda)| < 1 \quad \text{for} \quad 0 \leqq r \leqq r_4, \qquad \lambda \in \Lambda_4, \qquad 0 \leqq \varphi \leqq 2\pi.$$

Since $r_p(\varphi, \lambda) \to 0$ as $\lambda \to 0$ there is a positive λ_5, $\lambda_5 \leqq \lambda_4$, such that for $\lambda \in \Lambda_5$

$$\frac{1}{1 + \dfrac{\gamma(r_p(\varphi, \lambda), \varphi, \lambda)}{\beta(\lambda)}} = 1 - \frac{\gamma(r_p(\varphi, \lambda), \varphi, \lambda)}{\beta(\lambda)}$$
$$+ \frac{\gamma^2(r_p(\varphi, \lambda), \varphi, \lambda)}{\beta^2(\lambda)} + \frac{\gamma^3(r_p(\varphi, \lambda), \varphi, \lambda)}{\beta^3(\lambda)} \frac{1}{1 + \dfrac{\gamma(r_p(\varphi, \lambda), \varphi, \lambda)}{\beta(\lambda)}}. \tag{5.10}$$

To estimate $T(\lambda)$ we use the representation

$$r_p(\varphi, \lambda) = r_p^*(\varphi, \lambda) + \Delta_2(\varphi, \lambda) \tag{5.11}$$

where

$$r_p^*(\varphi, \lambda) = v_1 \sqrt{\lambda} + (v_2 + v_1^2 q_1(\varphi)) \lambda \tag{5.12}$$

and where according to (4.21)

$$\|\Delta_2(., \lambda)\| \leqq \bar{\varrho}_1^{(2)}(\lambda) = O(\lambda^{3/2}). \tag{5.13}$$

Hence, there are a positive number λ_6, $\lambda_6 \leqq \lambda_5$, and monotone increasing functions ϱ_i, $i = 1, 2, 3$, $\varrho_i: \Lambda_6 \to \mathbb{R}^+$ with $\varrho_i(\lambda) = O(\lambda^{3/2})$ as $\lambda \to 0$ such that for $\lambda \in \lambda_6$

$$\left| \left(\frac{1}{\beta(\lambda)}\right)^4 \int_0^{2\pi} \frac{\gamma^3(r_p(\varphi, \lambda), \varphi, \lambda)}{1 + \dfrac{\gamma(r_p(\varphi, \lambda), \varphi, \lambda)}{\beta(\lambda)}} d\varphi \right| \leqq \varrho_3(\lambda) = O(\lambda^{3/2}), \tag{5.14}$$

$$\left(\frac{1}{\beta(\lambda)}\right)^3 \int_0^{2\pi} \gamma^2(r_p(\varphi,\lambda),\varphi,\lambda)\,d\varphi = \frac{v_1^2\lambda}{(\beta(0)^3} \int_0^{2\pi} l_2(\varphi,0)\,d\varphi + R_2(\varphi,\lambda), \tag{5.15}$$

$$\|R_2(.,\lambda)\| \leqq \varrho_2(\lambda) = O(\lambda^{3/2}); \tag{5.16}$$

$$\left(\frac{1}{\beta(\lambda)}\right)^2 \int_0^{2\pi} \gamma(r_p(\varphi,\lambda),\varphi,\lambda)\,d\varphi = \frac{v_1^2\lambda}{(\beta(0))^2} \int_0^{2\pi} q_1(\varphi)\,l_1(\varphi,0)\,d\varphi + R_1(\varphi,\lambda), \tag{5.17}$$

$$\|R_1(.,\lambda)\| \leqq \varrho_1(\lambda) = O(\lambda^{3/2}). \tag{5.18}$$

From (5.3) we get

$$T(\lambda) = \frac{2\pi}{\beta(\lambda)} - \frac{1}{\beta(\lambda)^2} \int_0^{2\pi} \gamma(r_p(\varphi,\lambda),\varphi,\lambda)\,d\varphi$$

$$+ \frac{1}{\beta(\lambda)^2} \int_0^{2\pi} \gamma^2(r_p(\varphi,\lambda),\varphi,\lambda)\,d\varphi - \left(\frac{1}{\beta(\lambda)}\right)^4 \int_0^{2\pi} \frac{\gamma^3(r_p(\varphi,\lambda),\varphi,\lambda)}{1 + \frac{\gamma(r_p(\varphi,\lambda),\varphi,\lambda)}{\beta(\lambda)}}\,d\varphi. \tag{5.19}$$

Introducing the function

$$T^*(\lambda) := \frac{2\pi}{\beta(0)} - \lambda\left[\frac{2\pi\beta'(0)}{\beta(0)} + \frac{v_1^2}{(\beta(0))^2} \int_0^{2\pi} q_1(\varphi)\,l_1(\varphi,0)\,d\varphi - \frac{v_1^2}{(\beta(0))^3} \int_0^{2\pi} l_2(\varphi,0)\,d\varphi\right] \tag{5.20}$$

and taking into account the estimations (5.14)–(5.18) we obtain from (5.19) the inclusion

$$|T(\lambda) - T^*(\lambda)| \leqq T_r(\lambda) \tag{5.21}$$

where

$$T_r(\lambda) = O(\lambda^{3/2}) \quad \text{as} \quad \lambda \to 0. \tag{5.22}$$

Consequently we have the result

Theorem 5.1: *Assume the hypotheses* (V_1)–(V_4) *hold. Then there are a sufficiently small positive number* $\tilde{\lambda}$, $\tilde{\lambda} < \lambda^*$, *and a monotone increasing function* $T_r\colon [0,\tilde{\lambda}] \to \mathbb{R}^+$ *satisfying* (5.22) *such that the inclusion* (5.21) *is valid.*

Appendix

For convenience of the reader we give a formulation of the Newton-Kantorovič-Theorem [6] for equations depending on a parameter.

Proposition A: *Let* Ω *be an open subset of a Banach space* Z *with norm* $\|\,.\,\|$, *let* M *be a metric space, let* Q *be an operator mapping* $\Omega \times M$ *into* Z. *Suppose the following hypotheses:*

(H_1). $Q \in C^{2;0}_{x,\mu}(\Omega \times M, Z)$.

(H_2). *There are functions* $\bar{x}\colon M \to \Omega$, $\delta_0, k_1, k_2, k_3\colon M \to R^+$ *such that the following relations are valid for* $\mu \in M$,

$$\|Q(\bar{x}(\mu,\mu)\| \leqq k_1(\mu), \qquad \left\|\left[\frac{\partial Q}{\partial x}(\bar{x}(\mu),\mu)\right]^{-1}\right\| \leqq k_2(\mu), \tag{A.1}$$

$$\left\|\left[\frac{\partial^2 Q}{\partial x^2}(\tilde{x},\mu)\right]\right\| \leqq k_3(\mu) \quad \textit{for all} \quad \tilde{x} \in S_{\delta_0(\mu)} := \{x \in X\colon \|x - \bar{x}(\mu)\| < \delta_0(\mu)\},$$

$$k_1(\mu)\,k_2^2(\mu)\,k_3(\mu) < \tfrac{1}{2}, \qquad S_{\delta_1(\mu)} \subset S_{\delta_0(\mu)} \subset \Omega \tag{A.2}$$

where

$$\delta_1(\mu) = \frac{1 - \{1 - 2k_1(\mu)\,k_2^2(\mu)\,k_3(\mu)\}^{1/2}}{k_2(\mu)\,k_3(\mu)}. \tag{A.3}$$

Then to any $\mu \in M$ *the equation* $Q(x,\mu) = 0$ *has a solution in* $\overline{S_{\delta_1(\mu)}}$ *and this solution is unique in* $S_{\delta_0(\mu)} \cap S_{\delta_2(\mu)}$ *where*

$$\delta_2(\mu) = \frac{1 + \{1 - 2k_1(\mu)\,k_2^2(\mu)\,k_3(\mu)\}^{1/2}}{k_2(\mu)\,k_3(\mu)}. \tag{A.4}$$

References

1 CARR, J.: Application of centre manifold theory. Appl. Math. Sci. 35, Springer Verlag New York, Heidelberg, Berlin 1981.
2 FIEDLER, B.: Global Hopf bifurcation of two-parameter flows. Arch. Rat. Mech. Anal. **94** (1986), 59–81.
3 GOLUBITZKY, M.: A classification of degenerate Hopf bifurcation with $O(2)$ symmetry. J. Diff. Equs. **69** (1987), 216–264.
4 HASSARD, B. D.; KAZARINOFF, N. D.; WAN, Y.-H.: Theory and applications of Hopf bifurcation. London Math. Soc. Lect. Note Ser., Vol. **41**, Cambrigde 1981.
5 HOPF, E.: Abzweigung einer periodischen Lösung von einer stationären Lösung eines Differentialsystems. S.-Ber. Sächs. Akad. Wiss., Math.-Phys. Kl. **94** (1942), 1–22.
6 KANTOROWITSCH, L. W.; AKILOV, G. P.: Funktionalanalysis in normierten Räumen. Berlin 1964.
7 KNOBLOCH, H.-W.: Hopf bifurcation via integral manifolds. Abh. AdW der DDR N 3, 1977, Akademie-Verlag Berlin, pp. 413–429.
8 KOZYAKIN, V. S.; KRASNOSEL'SKII, M. A.: The method of parameter functionalization in the Hopf bifurcation problem. Nonlin. Anal. Meth. Appl. **11** (1987), 149–161.
9 LAUTERBACH, R.: Hopf bifurcation at a degenerate stationary pitchfork. Nonlin. Anal. Meth. Appl. **10** (1986), 339–351.
10 MARSDEN, J. E.; MCCRACKEN, M.: The Hopf bifurcation and its applications. Appl. Math. Sci., Vol. **19**, New York 1976.
11 SCHNEIDER, K. R.: Andronov-Hopf-Bifurkation im Falle stetiger Parameterabhängigkeit. ZAMM **60** (1980), 19–24.
12 SCHNEIDER, K. R.: Über die Anwendung der Mittelungsmethode zur Untersuchung der Abzweigung einer periodischen Lösung von einem singulären Punkt. In: Untersuchungen zur asymptotischen Theorie nichtlinearer Differentialgleichungen. Schriftenreihe Zentralinst. für Mathematik und Mechanik, Heft 26, 1978, pp. 57–64.
13 SCHNEIDER, K. R.; HEINZ, G.: Branch inclusion in generic Hopf bifurcation: The case of scaled van der Pol's equation. Internat. J. Nonlin. Mech. **26** (1991), 187–198.
14 SCHNEIDER, K. R.; WEGNER, B.: Hopf bifurcation of singularly perturbed differential systems. Math. Nachr. **112** (1983), 5–18.
15 VANDERBAUWHEDE, A.: Centre manifolds, normal forms and elementary bifurcations. Dynamics Reported **2** (1989), 89–169.
16 VANDERBAUWHEDE, A.; IOOS, G.: Center manifold theory in infinite dimensions. Preprint, Univ. Nice 1989.

Received April 9, 1990, revised January 10, 1991

Address: Dr. habil. KLAUS R. SCHNEIDER, Dr. G. HEINZ, Institut für Mathematik, Mohrenstraße 39, O-1086 Berlin-Germany

ZAMM · Z. angew. Math. Mech. 71 (1991) 7/8, 251–256 Akademie Verlag

REALI, M.; DASSIE, G.

Enhanced Accuracy and Applicability of Analytico-Numerical Solutions to Ordinary Differential Problems

Dedicated to the memory of LOTHAR COLLATZ

Der vorliegende Report beschäftigt sich mit den Möglichkeiten analytisch-numerischer Verfahren zur Lösung von Klassen gewöhnlicher Differentialprobleme, die auf Taylor-Reihenentwicklungen basieren und bemerkenswerte numerische Schwierigkeiten aufweisen. Derartige Verfahren, die relativ flexibel und leicht programmierbar sind, führen zu sehr genauen Lösungen, wie dies an Hand zweier im Detail ausgearbeiteter Aufgaben demonstriert wird. Die erste Aufgabe modelliert ein streng nichtlineares Ereignis, eine Unterwasserexplosion, die zu sphärischen Blasenschwingungen führt, die durch ultraschnelle Veränderungen der Blasen-Wasser-Phasengrenzflächen-Geschwindigkeit charakterisiert sind. Die zweite Aufgabe, eine stationäre Konvektions-Diffusions-Rechnung, wird ebenfalls auf sehr genaue Weise sogar für sehr große Peclet-Zahlen gelöst, wogegen Lösungen, die aus spezifischen Diskretisierungsverfahren erhalten wurden, wesentlich weniger genau sind.

This report concerns the potential of analytico-numerical techniques based on Taylor series developments for solving classes of ordinary differential problems presenting remarkable numerical difficulties. Such techniques which are rather flexible and easily programmable may lead to very accurate solutions as is demonstrated with two problems which are worked out in detail. The first problem models a strongly non-linear event, an underwater explosion leading to spherical bubble oscillations characterized by ultrarapid variations of the bubble/water interface velocity. The second problem, a steady-state convective-diffusion calculation, is also solved in a very accurate way even for very large Peclet numbers when solutions obtained from specific discretization techniques are much less accurate.

Этот доклад занимается с возможностью основанных на разложениях в теилоров ряд аналитико-численных способов для решения классов обыкновенных дифференциальных задач имеющих значительные численные трудности. Такие способы, которые являются довольно гибкими и легко программуроваными, ведут к очень точным решениям, как показыванно с помощью двух задач разработанных в деталь. Первая задача моделироваeт строго нелинейное событие: подводный взрыв ведущий к характеризированным при помощи сверхбыстрых изменений поверхностной скорости между пузырей и водой сферическим пузырковым колебаниям. Вторая задача, стационарное конвекционно-диффузионное вычисление, тоже решается очень точно даже для очень больших чисел Пекле, если решения полученных с помощью особенных методов дискретизации являются очень меньуие точными.

MSC (1980): 65L99

1. Introduction

Analytico- numerical techniques based on Taylor series developments (TST) have a long historical record among specialized methods for solving ordinary differential problems (ODP) [1–5].

We shall limit the discussion to ODP governed by one second-order ordinary differential equation (ODE) for the sake of simplicity. TST may be specially applied, eventually with the help of suitable functional transformations, to initial-value ODP in which the second derivative $\ddot{y}(x)$ of the unknown function $y(x)$ occurs explicitly and linearly. In the early versions of TST the required values of high-order derivatives were computed in terms of the initial conditions after deriving, recursively, the starting ODE. The considerable algebraic difficulties encountered made this procedure of rather limited applicability. However, for many classes of ODP, independent researchers (among whom the present authors) found that the Leibnitz rule for the derivative of the product of two functions can be a very powerful tool making the application of TST straightforward. Indeed, the computation of truncated Taylor series having a large number of terms and, consequently, a great accuracy is made possible through rather simple, easily programmable algebraic operations [6–8].

In section 2 we present an efficient application of TST to a specific initial-value ODP in bubble dynamics featuring remarkable numerical difficulties. We think it of interest to show that bubble dynamics problems can be effectively treated by TST since only Runge-Kutta type techniques appear to have been applied in the published literature [9–10]. Furthermore, we are not aware of any numerical solution, other than by the present TST, for the chosen problem. In view of the fact that most initial-value ODP are generally difficult to solve, the availability of two or more independent numerical-solution techniques should be of definite interest.

TST may also be considered for the treatment of linear two-point boundary-value ODP since the required solution can be found by superposing the solutions of two, properly associated, initial-value ODP. In section 3, a boundary-value ODP concerning a steady-state convective-diffusion calculation is solved by means of TST. Although this technique cannot be applied directly because of overflow problems, it will be shown that remarkably accurate solutions can be obtained, even in the case of very large Peclet numbers, by means of a simple numerical strategy.

The solution of non-linear boundary-value ODP is generally a difficult matter. Here we may notice that TST may be applied in a relatively effective way whenever shooting type techniques would be suitable.

2. An underwater explosion

An underwater explosion may be simulated in a simplified way by assuming that it gives rise to a spherical gas-bubble oscillation in an infinite medium free from boundaries and gravitational effects.

The following ODP describes the overall process:

$$R\ddot{R} + \tfrac{3}{2}\dot{R}^2 - AR^{-4} + B = 0\,, \tag{1.0}$$

$$R(t = t_0) = R_0, \qquad \dot{R}(t = t_0) = \dot{R}_0\,, \tag{1.1, 2}$$

where $R = R(t)$ represents the radius of the gas sphere at the time instant t, while $\dot{R} = \dot{R}(t)$ and $\ddot{R} = \ddot{R}(t)$ represent, respectively, the velocity and the acceleration of the gas-liquid interface. The parameters A and B represent given constants; $A = P_0 R_0^4 \varrho^{-1}$ where P_0 is the initial gas pressure, R_0 the initial bubble radius, and ϱ the water density; $B = P_D \varrho^{-1}$ where P_D is the depth pressure. The dimensions of A are $(\text{length})^6(\text{sec})^{-2}$ while those of B are $(\text{velocity})^2$.

In reference [11] problem (1.0-1-2) is solved analytically by means of elliptic integrals. A graphical representation of the solution is given in Fig. 1.

2.1 Analytico-numerical solution

Equation (1.0) is rewritten as follows,

$$R^5\ddot{R} + \tfrac{3}{2}R^4\dot{R}^2 - A + BR^4 = 0\,, \tag{2.0}$$

in order to make use of the following Taylor developments and relationships for the sought-for variables at the arbitrary time instants t_k and t_{k+1}, $k = 0, 1, 2, \ldots$:

$$R_{k+1} = \sum_{n=0}^{\infty} A_{n,k}\,, \qquad A_{n,k} = a_{n,k}\frac{(t_{k+1} - t_k)^n}{n!}\,; \qquad a_{n,k} = \frac{d^nR(t_k)}{dt^n}\,; \tag{3.0, 1}$$

$$\dot{R}_{k+1} = \sum_{n=0}^{\infty} B_{n,k}\,, \qquad B_{n,k} = A_{n+1,k}\frac{(n+1)}{(t_{k+1} - t_k)}\,; \tag{4.0, 1}$$

$$\ddot{R}_{k+1} = \sum_{n=0}^{\infty} C_{n,k}\,, \qquad C_{n,k} = B_{n+1,k}\frac{(n+1)}{(t_{k+1} - t_k)}\,; \tag{5.0, 1}$$

$$R_{k+1}^2 = \sum_{n=0}^{\infty} D_{n,k}\,, \qquad D_{n,k} = \sum_{r=0}^{n} A_{r,k}A_{n-r,k}\,; \tag{6.0, 1}$$

$$R_{k+1}^4 = \sum_{n=0}^{\infty} E_{n,k}\,, \qquad E_{n,k} = \sum_{r=0}^{n} D_{r,k}D_{n-r,k}\,; \tag{7.0, 1}$$

$$R_{k+1}^5 = \sum_{n=0}^{\infty} F_{n,k}\,, \qquad F_{n,k} = \sum_{r=0}^{n} A_{r,k}E_{n-r,k}\,; \tag{8.0, 1}$$

$$(\dot{R}_{k+1})^2 = \sum_{n=0}^{\infty} G_{n,k}\,, \qquad G_{n,k} = \sum_{r=0}^{n} B_{r,k}B_{n-r,k}\,; \tag{9.0, 1}$$

$$R_{k+1}^5\ddot{R}_{k+1} = \sum_{n=0}^{\infty} H_{n,k}\,, \qquad H_{n,k} = \sum_{r=0}^{n} F_{r,k}C_{n-r,k}\,; \tag{10.0, 1}$$

$$R_{k+1}^4(\dot{R}_{k+1})^2 = \sum_{n=0}^{\infty} P_{n,k}\,, \qquad P_{n,k} = \sum_{r=0}^{n} E_{r,k}G_{n-r,k}\,. \tag{11.0, 1}$$

By substituting positions (10.0), (11.0), and (7.0) into equation (2.0), the following fundamental relationship is obtained:

$$\sum_{n=0}^{\infty} H_{n,k} + \tfrac{3}{2}\sum_{n=0}^{\infty} P_{n,k} - A + B\sum_{n=0}^{\infty} E_{n,k} = 0 \tag{12.0}$$

while the initial conditions (1.1) and (1.2) become:

$$A_{0,k} = R_k\,, \qquad B_{0,k} = \dot{R}_k\,, \qquad A_{1,k} = B_{0,k}(t_{k+1} - t_k)\,. \qquad (13.0), (14.0), (15.0)$$

The required chain relationships, connecting all quantities of interest, are obtained according to standard procedure, from relation (12.0).

$n = 0$

Starting from:

$$H_{0,k} + \tfrac{3}{2} P_{0,k} - A + BE_{0,k} = 0 \tag{16.0}$$

one computes

$$D_{0,k} = A_{0,k}A_{0,k}, \quad E_{0,k} = D_{0,k}D_{0,k}, \quad F_{0,k} = A_{0,k}E_{0,k}, \quad G_{0,k} = B_{0,k}B_{0,k}, \qquad (16.1, 2, 3, 4)$$

$$P_{0,k} = E_{0,k}G_{0,k}, \quad C_{0,k} = F_{0,k}^{-1}[A - \tfrac{3}{2} P_{0,k} - BE_{0,k}], \qquad (16.5, 6)$$

$$B_{1,k} = C_{0,k}(t_{k+1} - t_k). \quad A_{2,k} = B_{1,k} \frac{(t_{k+1} - t_k)}{2}. \qquad (16.7, 8)$$

$n = n \qquad n = 1, 2, \ldots$

Starting from:

$$H_{n,k} + \tfrac{3}{2} P_{n,k} + BE_{n,k} = 0 \tag{17.0}$$

one computes

$$D_{n,k} = \sum_{r=0}^{n} A_{r,k} A_{n-r,k}, \quad E_{n,k} = \sum_{r=0}^{n} D_{r,k} D_{n-r,k}, \qquad (17.1, 2)$$

$$F_{n,k} = \sum_{r=0}^{n} A_{r,k} E_{n-r,k}, \quad G_{n,k} = \sum_{r=0}^{n} B_{r,k} B_{n-r,k}, \qquad (17.3, 4)$$

$$P_{n,k} = \sum_{r=0}^{n} E_{r,k} G_{n-r,k}, \quad C_{n,k} = -F_{0,k}^{-1}\left[\tfrac{3}{2} P_{n,k} + BE_{n,k} + \sum_{r=1}^{n} F_{r,k} C_{n-r,k}\right], \qquad (17.5, 6)$$

$$B_{n+1,k} = C_{n,k}(t_{k+1} - t_k), \quad A_{n+2,k} = B_{n+1,k} \frac{(t_{k+1} - t_k)}{2}. \qquad (17.7, 8)$$

The sought-for solution $R(t_{k+1}) = R_{k+1}$ and its first derivative $\dot{R}_{k+1}$ are computed from the truncated Taylor series:

$$R_{k+1} = \sum_{n=0}^{N_k} A_{n,k}, \quad \dot{R}_{k+1} = \sum_{n=0}^{N_k - 1} B_{n,k}, \qquad (18.0, 1)$$

where N_k is an arbitrary natural integer and $\Delta t_k = t_{k+1} - t_k$ is a time interval which is kept sufficiently small for meeting fixed precision requirements. In the actual computations N_k has been kept equal to 19 for any k value and Δt_k has been kept small enough to make $|A_{19,k}| < 10^{-30}$.

Computations made in double precision on a VAX 6310 are represented graphically in Fig. 1 on the basis of the following numerical data:

$$R_0 = 1, \quad \dot{R}_0 = 0; \quad A = 10^3; \quad B = 10^5.$$

As for Δt_k, the following values have been utilized: 3.0 10^{-6}, 1.0 10^{-6}, 3.0 10^{-7}, 1.0 10^{-7}, 3.0 10^{-8}, 1.0 10^{-8}, 3.0 10^{-9}.

3. One-dimensional steady convective diffusion

The steady-state convective-diffusion problem

$$\ddot{u} - \mathrm{Pe}\,\dot{u} = 0, \quad 0 \leqq x \leqq 1, \quad u(x = 0) = 0, \quad u(x = 1) = 1, \qquad (19.0)\ (19.1, 2)$$

where u represents a temperature difference and Pe is the so-called Peclet number, has been utilized for evaluating several numerical methods [12].

From the graphical representation, in Fig. 2, of the analytic solution

$$u = [1 - \exp(\mathrm{Pe})]^{-1} [1 - \exp(\mathrm{Pe}\, x)] \tag{20.0}$$

one can foresee that solving the problem numerically for large Peclet numbers would entail considerable difficulties.

We have found that TST can be applied to problem (19.0-1-2) in an efficient and straightforward way to reach an analytico-numerical solution which is much more accurate than any numerical solution analyzed in [12].

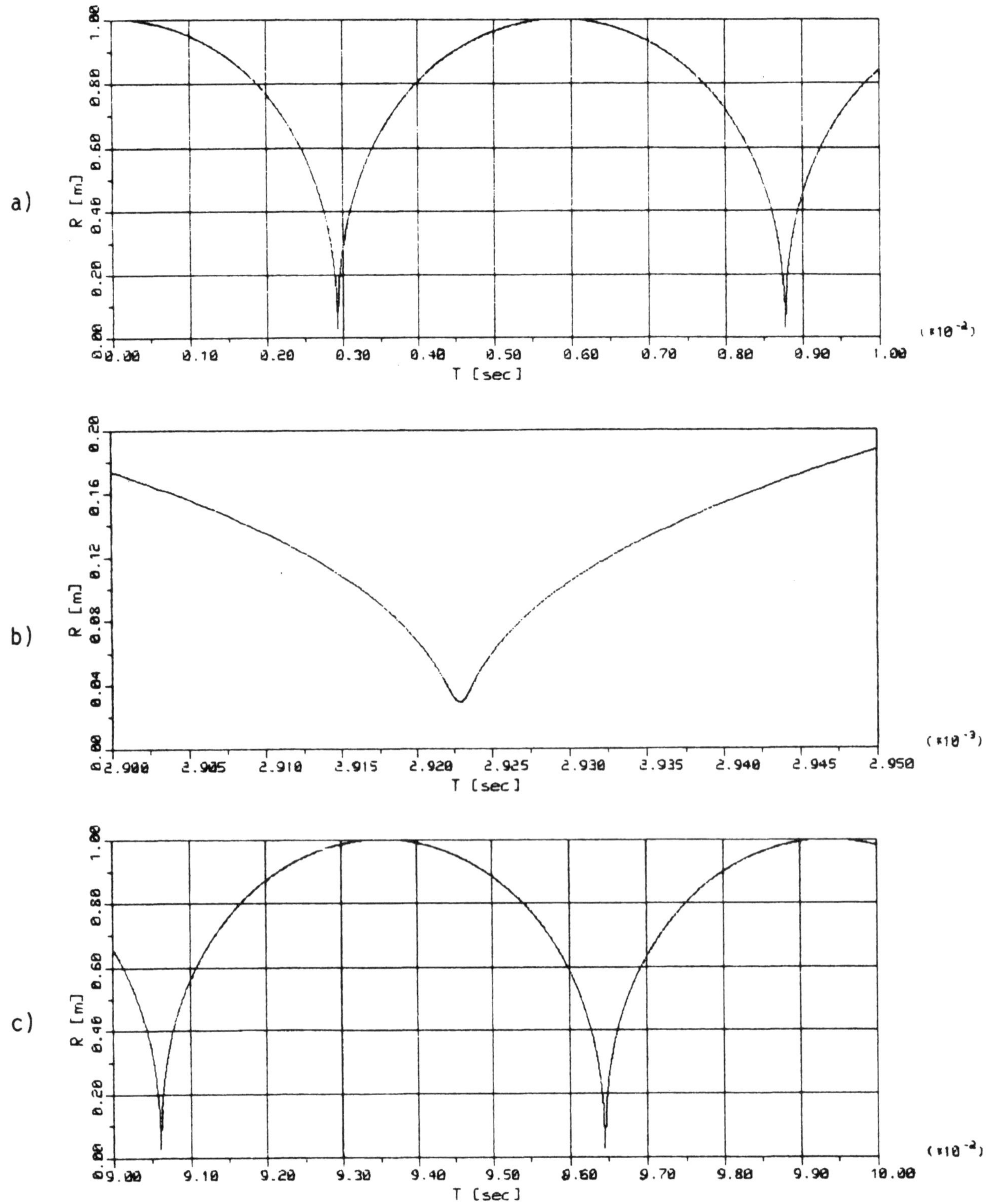

Fig. 1. Graphical representation of the present analytico-numerical solution to problem (1.0-1-2) – a) representation for the interval $0 \leqq t \leqq 1.0\ 10^{-2}$; – b) enlargement about the first minimum of the radius R for the interval $2.9\ 10^{-3} \leqq t \leqq 2.95\ 10^{-3}$; – c) representation for the interval $9.0\ 10^{-2} \leqq t \leqq 1.0\ 10^{-1}$: the spherical bubble oscillations appear to be stable. This graphical representation appears to be almost equivalent to that corresponding to the analytical solution of Reference 11

Specifically, in order to cope with overflow difficulties (in double precision computations on a VAX 6310) we have considered two variants of TST: the first one makes use of a logarithmic dependent variable transformation, while the second one introduces local similarity transformations. Here, we shall describe only the second variant.

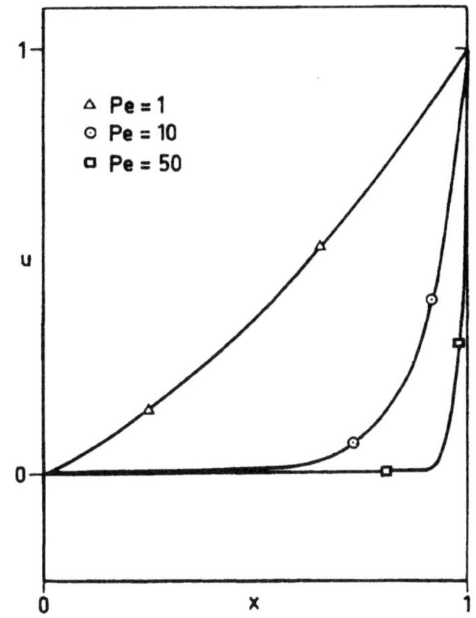

Fig. 2. Graphical representation of the analytic solution (20.0) to problem (19.0-1-2)

3.1 Analytico-numerical solution with local similarity transformations

If the points x_i, $i = 0, 1, 2 \dots N$; $x_0 = 0$; $x_N = 1$, denote an arbitrary subdivision of the interval $0 \leqq x \leqq 1$, we introduce the following similarity transformations:

$$u(x) = ay(x), \qquad 0 \leqq x \leqq 1, \tag{21.0}$$

$$y(x) = y_{0,1}(x), \qquad 0 \leqq x \leqq x_1, \tag{22.0}$$

$$y(x) = \prod_{k=1}^{i} a_k y_{i,i+1}(x), \qquad x_i \leqq x \leqq x_{i+1}, \qquad i = 1, 2, \dots, N-1, \tag{23.0}$$

where the coefficients a_k are to be determined in the computation of the functions $y_{k,k+1}$ and the coefficient a is to be determined from the boundary condition (19.2).

Function $y_{0,1}$ satisfies the equation

$$\ddot{y}_{0,1} - \mathrm{Pe}\,\dot{y}_{0,1} = 0 \tag{24.0}$$

and the initial conditions:

$$y_{0,1}(x = 0) = 0, \qquad \dot{y}_{0,1}(x = 0) = b \tag{24.1, 2}$$

where b is an arbitrary sufficiently small number. Also the computational interval $[0, x_1]$ is to be sufficently small.

Function $y_{1,2}$ is the solution to the initial value ODP:

$$\ddot{y}_{1,2} - \mathrm{Pe}\,\dot{y}_{1,2} = 0, \qquad x_1 \leqq x \leqq x_2, \tag{25.0}$$

$$y_{1,2}(x_1) = (a_1)^{-1}\, y_{0,1}(x_1), \qquad a_1 = b^{-1}\dot{y}_{0,1}(x_1); \qquad \dot{y}_{1,2}(x_1) = b. \tag{25.1, 2}$$

Similarly, function $y_{i,i+1}$ is the solution to the initial value ODP:

$$\ddot{y}_{i,i+1} - \mathrm{Pe}\,\dot{y}_{i,i+1} = 0, \qquad x_i \leqq x \leqq x_{i+1}, \qquad i = 2, 3, \dots, N-1, \tag{26.0}$$

$$y_{i,i+1}(x_i) = (a_i)^{-1}\, y_{i-1,i}(x_i), \qquad a_i = b^{-1}\dot{y}_{i-1,i}(x_i); \qquad \dot{y}_{i,i+1}(x_i) = b. \tag{26.1, 2}$$

Once all functions $y_{i,i+1}$, $i = 0, 1, 2 \dots N-1$, have been computed, one can determine the coefficient a in (21.0) which defines the required solution u. One obtains:

$$u(1) = ay(1), \tag{27.0}$$

$$1 = a \prod_{k=1}^{N-1} a_k y_{N-1,N}(1), \qquad a = \prod_{k=1}^{N-1} (a_k)^{-1}\,[y_{N-1,N}(1)]^{-1}. \tag{27.1, 2}$$

Numerical results, corresponding to the constant value $b = 10^{-20}$ and to a constant number (20) of terms in the truncated Taylor series, with the absolute value of the last term being kept smaller than 10^{-30} are collected in table 1 where u^* denotes the present analytico-numerical solution and $(\Delta x)_{\min}$ the smallest step utilized in the Taylor series.

Table 1. Numerical results for problem (19.0-1-2) obtained with the present TST in association with local similarity transformations.

Pe	10^2	10^3	10^4	10^5
N	5	50	400	4000
$x_{i+1} - x_i$, $i = 0, 1, \dots, N-1$	$2.0\ 10^{-1}$	$2.0\ 10^{-2}$	$2.5\ 10^{-3}$	$2.5\ 10^{-4}$
$(\Delta x)_{\min}$	10^{-2}	10^{-3}	10^{-4}	10^{-5}
a_1	$4.8\ 10^8$	$4.8\ 10^8$	$7.2\ 10^{10}$	$7.2\ 10^{10}$
$y_{1,2}(x_2)$	$4.8\ 10^{-14}$	$4.8\ 10^{-15}$	$7.2\ 10^{-14}$	$7.2\ 10^{-15}$
$\max \lvert u - u^* \rvert$	$1.7\ 10^{-16}$	$1.3\ 10^{-16}$	$3.9\ 10^{-15}$	$5.2\ 10^{-14}$

We notice that all a_k, $k = 1, 2, \dots, N-1$, have practically the same value, and analogously for all $y_{k,k+1}(x_{k+1})$. Consequently, only the values of a_1 and of $y_{1,2}(x_2)$ are given in Table 1.

As a final consideration we mention that the maximum errors obtained through the various finite-difference methods analyzed in [12] are several orders of magnitude larger than the maximum error obtained with the present TST.

4. Conclusive comments

As has been shown in sections 2 and 3, TST is a direct, noniterative, self-starting, one-step analytico-numerical technique which is conceptually simple, has a wide range of applicability, and is easily programmable on an electronic computer. TST as an independent, generally highly accurate numerical technique, should help to provide deep insight into the varied and oftentimes extraordinary phenomena described by non linear initial value ODP.

References

1 Collatz, L.: The numerical treatment of differential equations. Springer Verlag, Berlin 1966.
2 Milne, W. E.: Numerical solution of differential equations. Dover Pub. Inc., New York 1970.
3 Henrici, P.: Discrete variable methods in ordinary differential equations. John Wiley & Sons, Inc., New York 1962.
4 Davis, H. T.: Introduction to nonlinear differential and integral equations. Dover Pub. Inc., New York 1962.
5 Lambert, J. D.: Computational methods in ordinary differential equations. John Wiley & Sons, London 1973.
6 Barton, D.; Willers, I. M.; Zahar, R. V. M.: Taylor series methods for ordinary differential equations – an evaluation. In Rice, J. R. (ed): Mathematical software. Academic Press, New York 1971.
7 Corliss, G.; Chang, Y. F.: Solving ordinary differential equations using Taylor series. ACM Trans. Math. Software **8** (1982) 2, 114–144.
8 Reali, M.; Dassie, G.: A Taylor series expansion technique for initial-value non linear ordinary differential problems. In: Proc. 11th Internat. Conf. on Nonlinear Oscillations – Jànos Bolyai Mathematical Society, Budapest 1987.
9 Lauterborn, W.: Numerical investigation of nonlinear oscillation of gas bubbles in liquids. J. Acoust. Soc. Amer. **59** (1976), 283–293.
10 Flynn, H. G.: Transient pulsations of small gas bubbles in water. J. Acoust. Soc. Amer. **84** (1988), 985–998.
11 Childs, D. R.: Exact solution of the non linear differential equation $R\ddot{R} + 3/2\dot{R}^2 - AR^{-4} + B = 0$. Internat. J. Non-linear Mech. **8** (1973), 371–379.
12 Gupta, M. M.: A survey of some second-order difference schemes for the steady-state convection-diffusion equation. Internat. J. Numer. Meth. in Fluids **3** (1983), 319–331.

Address: Dr. Mario Reali, Mr. Giuseppe Dassie, ENEL DSR-CRIS, Via Ornato, 90/14, 20162 Milano, Italy

ZAMM · Z. angew. Math. Mech. 71 (1991) 7/8, 257–265 Akademie Verlag

AL-FAHED, A. M.; STAVROULAKIS, G. E.; PANAGIOTOPOULOS, P. D.

Hard and Soft Fingered Robot Grippers. The Linear Complementarity Approach

Dedicated to the memory of LOTHAR COLLATZ

Die vorliegende Arbeit beschäftigt sich mit dem Problem, Werkstücke mit einem vielfingerigen Reibungsgreifer aufzugreifen. Dabei werden zwei Typen von Reibungsfingern behandelt, „harte" und „sanfte" Finger. Harte Finger haben die Fähigkeit, die Verschiebungen des Objektes auf normale und tangentiale Richtungen bezogen auf die Objektoberfläche zu beschränken, während die sanften Finger zusätzlich die Fähigkeit haben, die Drehung des Objektes auf die Rotation um die Normalachse durch den Kontaktpunkt einzuschränken. Sowohl das einseitige Kontaktproblem als auch das Reibungsproblem werden in kompakter Form als unsymmetrisches lineares Komplementaritätsproblem formuliert. Notwendige und hinreichende Bedingungen, die die Lösbarkeit des Problems sichern, werden abgeleitet. Ebenso werden numerische Beispiele angegeben, in denen die Normal- und Tangentialreaktion, die Lagrangeschen Gleitmultiplikatoren und die Verschiebungen und Rotationen des Festkörpers berechnet werden.

This paper deals with the problem of grasping work pieces with frictional multifingered grippers. The two types of frictional fingers "hard finger" and "soft finger" are treated. Hard fingers have the ability to constraint the object displacements on the normal and tangential directions to the object boundary, whereas soft fingers have in addition the ability to constraint the rotation of the object about the normal axis through the point of contact. The unilateral contact problem and the frictional problem are formulated in a compact form as a nonsymmetric Linear Complementarity Problem (LCP). Necessary and sufficient conditions which guarantee the solvability of the problem are derived. Also numerical examples computing the normal and tangential reactions, the slip Lagrange multiplier and the rigid body displacements and rotations are given.

Статья занимается с задачей схватывания заготовок с помощью грейфера с много пальцев с трением. Учитываются две рода пальцев с трением: «жесткие пальцы» и «мяские палбцы». Жесткие пальцы могут огранчить смещения об'екта на нормальное и на тангенциальное направления относительно границы об'екта, поскольку мягкие пальцы кроме того могут ограничить вращение об'екта вдоль нормальной оси через контактную точку. Формулируются одностороннюю контактную задачу и задачу с трением в компактной форме в виде несимметричной линейной дополнительной задачи. Введутся необходимые и достаточные условия для разрешимости задачи. Кроме того даются численные примеры для вычисленния нормальных и тангенциальных реакций, лагранжевые множители скольжения как и смещения и вращения твердого тела.

MCS (1980): 70E99

1. Introduction

The interest in multifingered dexterous grippers has been growing in robotics during the last years [1]–[5]. Force control of the gripper-object system requires the specification of the contact forces which exert the fingertips on the object. The interaction between the gripper and the object exhibits variable contact conditions, where the reactions of the fingertips are indeterminate and depend upon the external applied forces.

These types of problems are nonlinear and their difficulties lie in the fact that the finger which will be in contact with the object (or the contact region on the object surface) is not known a priori. The behaviour of the system is constrained first, by the inequality restriction which is introduced by the fact that the fingertip and the object can only push to each other and not pull (in other words the system is similar to the interaction of legged locomotion systems with the ground), and Coulomb's law. In recent years many researchers have studied problems with unilateral frictional phenomena [6]–[12]. PANAGIOTOPOULOS suggests a method to treat contact problems with friction consisting in the decomposition of the problem into one unilateral contact problem with known friction forces, and one with known contact forces. Finally a fixed-point-type iterative technique is used for the numerical solution of the problem.

Firm grips of an object by a frictional multifingered gripper (both the so-called cases of "form closure" and "force closure" under friction) have been recently investigated by many researchers [2]–[4], [13]. Several of them have studied the problem by satisfying only the equilibrium equations and constraints introduced by the friction law with the assumption of nonnegative normal reactions [3], [4] or by considering the special features of the object geometry [13]. In all the aformentioned methods both the fingers and object are assumend to be rigid.

A stable grasping and a fine manipulation depend strongly on the contact properties of the gripper-object system. "Hard-finger" contact and "soft-finger" [1] contact have different abilities in order to constrain the object. The hard-finger contact which actually is point contact with friction has the ability to prevent the displacements of the object in the normal direction as well as in the tangential direction up to the frictional limit; nevertheless rotation about the axis which is normal to the boundary of the object and passes through the point contact is free. On the other hand the soft-finger contact has in addition the ability to prevent the rotation of the object about the normal axis through the contact point up to a frictional limit. In general it is clear that owing to the friction the number of the fingers needed to grasp firmly an object is less than that in the absence of friction.

In this work, a mathematical programming approach to the grasping contact problems with friction, is presented. The global equilibrium equations are considered, and the unilateral normal conditions, as well as the unilateral friction conditions are presented for the two cases of hard-finger contact and soft-finger contact. Then the corresponding Linear Complementarity Problem (LCP) is formulated. The friction is assumed to obey Coulomb's law. The numerical treatment of the resulting LCP by an appropriate algorithm gives normal reaction, tangential forces (friction forces), the slipping when it occurs and the corresponding rigid body displacement. In this paper a variant of LEMKE's algorithm (c.f. [14], [15]) is used to solve the problem. Numerical examples illustrate the theory. It is also shown by means of a numerical example that an object which is grasped with soft-fingers cannot be grasped with hard-fingers for the same external forces.

2. Problem formulation

Let us consider in the orthogonal Cartesian coordinate system $O_{\chi 1 \chi 2 \chi 3}$ a rigid object which is grasped by a multifingered gripper with n frictional elastic fingers (Fig. 1a). The contact between the object and the fingertips is modeled either as a hard-finger contact (Fig. 1b) or as a soft-finger contact (Fig. 1c). The hard-finger contact can prevent displacements of the object in the normal and the tangential directions with respect to the object boundary, whereas the soft-finger contact

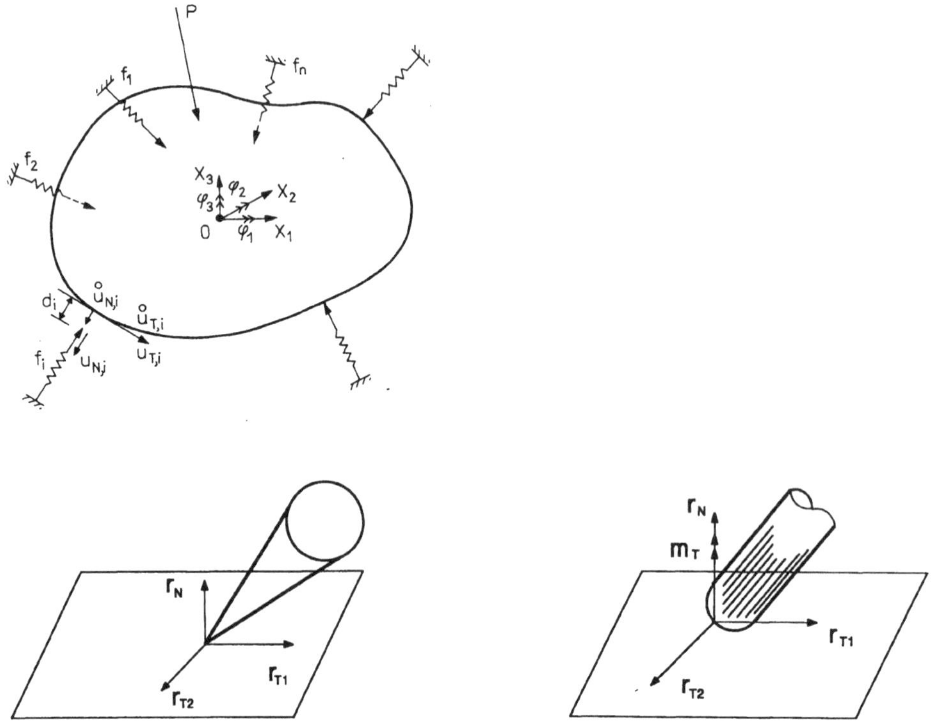

Fig. 1. a) Rigid object in an elastic frictional multifingered gripper. b) Hard finger-contact. c) Soft finger-contact

can prevent in addition rotations of the object about the normal to the boundary axis passing through the contact point, up to a certain value of torque which depends on the normal contact force through Coulomb-type static friction. Under the above assumptions the relations which govern the frictional contact problem are the following:

i) The global equilibrium equations

$$\boldsymbol{G}_N \boldsymbol{r}_N + \boldsymbol{G}_T \boldsymbol{r}_T = \boldsymbol{P}, \tag{1}$$

where $\boldsymbol{G}_N$ is the $m \times n$ equilibrium matrix corresponding to the normal contact reactions, $\boldsymbol{G}_T$ is the $m \times 2n$ ($m \times 3n$ for the soft-finger case) equilibrium matrix corresponding to the friction forces (plus the frictional torques in the soft-finger case), $\boldsymbol{r}_N = \{r_{N1}, r_{N2}, \ldots, r_{Nn}\}^T$ are the normal forces or reactions which exert the fingertips on the object, and $\boldsymbol{r}_T$ are the friction forces which are tangential to the boundary. For hard-finger contact $\boldsymbol{r}_T = \{r_{T11}, r_{T12}, r_{T21}, r_{T22}, \ldots, r_{Tn1}, r_{Tn2}\}^T$ whereas for soft-finger contact $\boldsymbol{r}_T = \{r_{T11}, r_{T12}, m_{T1}, r_{T21}, r_{T22}, m_{T2}, \ldots, r_{Tn1}, r_{Tn2}, m_{Tn}\}^T$. Here $\boldsymbol{m}_T$ is the moment about the surface normal, exerted by the finger onto the body, if one assumes soft-finger contact.

ii) The kinematical conditions.

On the normal direction to the surface of the body the compatibility conditions are written in the form:

$$\boldsymbol{u}_N + \boldsymbol{d}_N = \mathring{\boldsymbol{u}}_N , \tag{2}$$

where $\boldsymbol{u}_N$ is the vector of the normal displacements of the fingertips, $\mathring{\boldsymbol{u}}_N$ denotes the normal displacements of the points on the object boundary which are adjacent to the fingertips, and $\boldsymbol{d}_N$ is the vector of the initial openings between the fingertips and the object. The unilateral behaviour of the system can be expressed by the following unilateral contact conditions:

$$\text{if} \quad \boldsymbol{u}_N + \boldsymbol{d}_N = \mathring{\boldsymbol{u}}_N , \quad \text{then} \quad \boldsymbol{r}_N \geqq \boldsymbol{0}$$

(if contact occurs, then the reaction is nonnegative);

$$\text{if} \quad \boldsymbol{u}_N + \boldsymbol{d}_N > \mathring{\boldsymbol{u}}_N , \quad \text{then} \quad \boldsymbol{r}_N = \boldsymbol{0}$$

(if no contact occurs, then the reaction is zero).

The above relations can be written in the following linear complementarity form:

$$\boldsymbol{u}_N - \mathring{\boldsymbol{u}}_N + \boldsymbol{d}_N = \boldsymbol{y}_N , \tag{3}$$

$$\boldsymbol{y}_N \geqq \boldsymbol{0} , \qquad \boldsymbol{r}_N \geqq \boldsymbol{0} ; \qquad \boldsymbol{y}_N^T \boldsymbol{r}_N = 0 . \tag{4 a; b}$$

iii) Friction conditions.

Coulomb's law of dry friction implies that the tangential (frictional) forces, which exert the fingertips on the object must satisfy the following relation,

$$\gamma_i = \mu |r_{Ni}| - |r_{Ti}| , \qquad i = 1, \ldots, n ; \qquad \gamma_i \geqq 0 , \tag{5 a; b}$$

and in addition for soft-finger contact,

$$\gamma_i = \mu_t |r_{Ni}| - |m_{Ti}| , \qquad i = 1, \ldots, n ; \qquad \gamma_{iT} \geqq 0 , \tag{5 c; d}$$

where $|*|$ denotes the norm in R^3, and μ is the friction coefficient. If $|r_{Ti}| < \mu |r_{Ni}|$ (i.e. $\gamma_i > 0$) the slipping value y_{Ti} must equal to zero, and if $|r_{Ti}| = \mu |r_{Ni}|$ (i.e. $\gamma_i = 0$) then $y_{Ti} \neq 0$ and we have slipping in the opposite direction of r_{Ti}. In addition for the case of soft-finger contact if $|m_{Ti}| < \mu_t |r_{Ni}|$ (i.e. $\gamma_{iT} > 0$) the slipping value y_{TiT} must equal zero, and if $|m_{Ti}| = \mu_t |r_{Ni}|$ (i.e. $\gamma_{iT} = 0$) then the slipping value $y_{TiT} \neq 0$ and we have slipping in the opposite direction of m_{Ti}. Thus we may write [7]:

$$\text{If} \quad \gamma_i > 0 \quad \text{then} \quad y_{Ti} = 0 . \tag{6a}$$

$$\left.\begin{array}{l} \text{If} \quad \gamma_i = 0 \quad \text{then there exists } \sigma > 0 \\ \qquad \text{such that} \quad y_{Ti} = -\sigma r_{Ti} , \qquad \sigma \text{ unknown} . \end{array}\right\} \tag{6b}$$

In addition for the soft-finger case:

$$\text{If} \quad \gamma_{iT} > 0 \quad \text{then} \quad y_{TiT} = 0 . \tag{7a}$$

$$\left.\begin{array}{l} \text{If} \quad \gamma_{iT} = 0 \quad \text{then there exists } \sigma_T > 0 \\ \qquad \text{such that} \quad y_{TiT} = -\sigma_T m_{Ti} , \qquad \sigma_T \text{ unknown} , \end{array}\right\} \tag{7b}$$

which holds if we have contact, i.e. if $y_{Ni} = 0$.

3. Formulation of the Linear Complementarity Problem (LCP)

In this section we formulate the problem defined in the previous section by the relations (1)–(7) as a Linear Complementarity Problem (LCP). To achieve this we use a piecewise linearisation of the friction law (5) by a polyhedral approximation of the friction cone from the interior (Fig. 2). By means of this approximation the relation (5) can be written as follows:

$$\gamma = \boldsymbol{T}_N^T \boldsymbol{r}_N + \boldsymbol{T}_T^T \boldsymbol{r}_T . \tag{8}$$

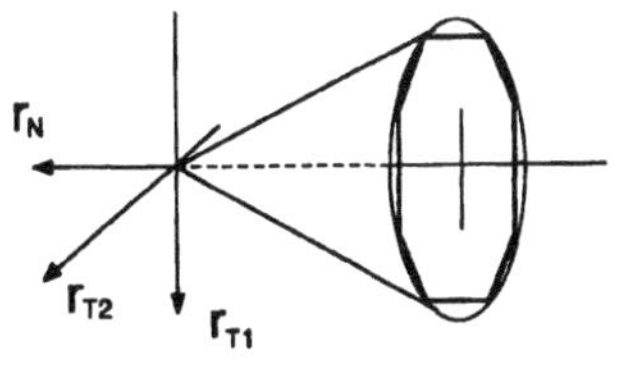

Fig. 2. Linearized friction cone

The superscript T denotes the transpose of a vector or a matrix. The matrices T_T and T_N of the linearised friction law have the following form:

$$T_T = \operatorname{diag}[T_T^1, T_T^2, \dots, T_T^n], \qquad T_N = \operatorname{diag}[T_N^1, T_N^2, \dots, T_N^n], \tag{9, 10}$$

where for hard-finger

$$T_T^i = \begin{bmatrix} \cos a_1 & \dots & \cos a_l \\ \sin a_2 & \dots & \sin a_l \end{bmatrix}, \qquad a_j = (j-1)\frac{2\pi}{l};$$

$$T_N^i = [\mu_1, \dots, \mu_l], \qquad \mu_j = \mu \cos\left(\frac{\pi}{l}\right)$$

and for soft-finger

$$T_T^i = \begin{bmatrix} \cos a_1 & \dots & \cos a_l & 0 & 0 \\ \sin a_1 & \dots & \sin a_l & 0 & 0 \\ 0 & & 0 & 1 & -1 \end{bmatrix}, \qquad a_j = (j-1)\frac{2\pi}{l};$$

$$T_N^i = [\mu_1, \dots, \mu_l, \mu_T, \mu_T], \qquad \mu_j = \mu \cos\left(\frac{\pi}{l}\right),$$

where l is the number of the surfaces of the linearised friction cone, μ is the coefficient of friction, and μ_T is the torsional friction coefficient. Numerical experiments have shown that the internal linearization of the friction cone leads to more accurate results than the external linearisation used in [16]. Linearisation of the friction cone does not lead to a best performance algorithm in the case of large scale discretized continua, where the consideration of the friction cone without linearisation [11] is preferable. Nevertheless, the latter approach necessitates the use of iterative methods for the numerical solution which is not the case in this paper. Due to (6), (7) the slip value can be written in the following form by taking appropriate projections:

$$y_T = T_T\lambda, \qquad \lambda \geqq 0, \tag{11}$$

where λ is a nonnegative slipping parameter-vector. Then y and λ fullfil the following orthogonality condition:

$$y^T\lambda = 0. \tag{12}$$

Note that the slipping value y_T, the tangential displacements of the fingertips u_T, and the tangential displacements $\mathring{u}_T$ of the adjacent points on the object surface due to rigid body motion, are related by the condition:

$$T_T\lambda - u_T + \mathring{u}_T = d_T, \tag{13}$$

where d_T denotes the initial tangential distance between the fingertip and the potential point of contact on the object boundary.

Assuming a linear behaviour of the fingers one can introduce the reaction-displacement law:

$$u = Fr, \tag{14}$$

where

$$u = \begin{bmatrix} u_N \\ u_T \end{bmatrix}, \qquad F = \begin{bmatrix} F_{NN} & F_{NT} \\ F_{TN} & F_{TT} \end{bmatrix}, \qquad r = \begin{bmatrix} r_N \\ r_T \end{bmatrix},$$

F is the symmetric matrix of flexibility and F_{NN} is an $n \times n$ nonsingular matrix and F_{TT} is a $2n \times 2n$ ($3n \times 3n$ for soft-finger contact) nonsingular matrix.

By means of the principle of complementary virtual work one can find that:

$$\mathring{u}_N = -G_N^T\mathring{u}, \qquad \mathring{u}_T = -G_T^T\mathring{u}, \tag{15), (16}$$

where $\mathring{u}$ is the vector of the rigid body displacements and rotations with respect to the coordinate system. For three-dimensional problems $\mathring{u}$ reads:

$$\mathring{u} = \{\mathring{u}_1, \mathring{u}_2, \mathring{u}_3, \mathring{\varphi}_1, \mathring{\varphi}_2, \mathring{\varphi}_3\}^T. \tag{17}$$

Here $\mathring{u}_1, \mathring{u}_2, \mathring{u}_3$ are the rigid body displacements along the directions O_{x1}, O_{x2}, O_{x3} respectively, and $\mathring{\varphi}_1, \mathring{\varphi}_2, \mathring{\varphi}_3$ denote the rigid body rotations about the axes of the coordinate system. By means of the relations (14), (15), and (16), (3) and (11) take the following form:

$$y_N - F_{NN}r_N - F_{NT}r_T - G_N^T\mathring{u} = d_N, \qquad T_T\lambda - F_{TN}r_N - F_{TT}r_T - G_T^T\mathring{u} = d_T. \tag{18), (19}$$

Let us now express the rigid body displacements as a difference of two nonnegative values (see [6] p. 352),

$$\mathring{u} = \mathring{u}^+ - \mathring{u}^-, \tag{20}$$

where

$$\mathring{u}^+ = \frac{\mathring{u} + |\mathring{u}|}{2} \geqq 0 \quad \text{and } \mathring{u}^- = \frac{-\mathring{u} + |\mathring{u}|}{2} \geqq 0.$$

Substituting (20) in (18) and (19) we have

$$y_N - F_{NN}r_N - F_{NT}r_T - G_N^T\mathring{u}^+ + G_N^T\mathring{u}^- = d_N, \tag{21}$$

$$T_T\lambda - F_{TN}r_N - F_{TT}r_T - G_T^T\mathring{u}^+ + G_T^T\mathring{u}^- = d_T. \tag{22}$$

Now the nonnegative slack vectors $v^+ \geqq 0$, $v^- \geqq 0$ are introduced in equation (1), which can be put in the form

$$v^+ + G_Nr_N + G_Tr_T = P, \qquad v^- - G_Nr_N - G_Tr_T = -P. \tag{23 a, b}$$

From equations (1), (23a) and (23b), we note that at the equilibrium state the nonnegative slack vectors v^+ and v^- must be equal to zero and thus the following holds:

$$(v^+)^T \mathring{u}^+ = 0, \qquad (v^-)^T \mathring{u}^- = 0. \tag{24}$$

Using equation (22), r_T can be expressed as follows:

$$r_T = -F_{TT}^{-1}F_{TN}r_N - F_{TT}^{-1}G_T^T\mathring{u}^+ + F_{TT}^{-1}G_T^T\mathring{u}^- + F_{TT}^{-1}T_T\lambda - F_{TT}^{-1}d_T. \tag{25}$$

In the numerical calculations which follow the inverse of F_{TT} is performed by the LDL^T decomposition.

By eliminating r_T from the equations (21), (23 a, b) and (8), we obtain

$$y_T - (F_{NN} - F_{NT}F_{TT}^{-1}F_{TN})\, r_N - (G_N^T - F_{NT}F_{TT}^{-1}G_T^T)\, \mathring{u}$$
$$+ (G_N^T - F_{NT}F_{TT}^{-1}G_T^T)\, \mathring{u}^- + F_{NT}F_{TT}^{-1}T_T\lambda = d_N - F_{NT}F_{TT}^{-1}d_T, \tag{26}$$

$$v^+ + (G_N - G_TF_{TT}^{-1}F_{TN})\, r_N - G_TF_{TT}^{-1}G_T^T\mathring{u}^+ + G_TF_{TT}^{-1}G_T^T\mathring{u}^-$$
$$+ G_TF_{TT}^{-1}T_T\lambda = P + G_TF_{TT}^{-1}d_T, \tag{27a}$$

$$v^+ - (G_N - G_TF_{TT}^{-1}F_{TN})\, r_N + G_TF_{TT}^{-1}G_T^T\mathring{u}^+ - G_TF_{TT}^{-1}G_T^T\mathring{u}^-$$
$$- G_TF_{TT}^{-1}T_T\lambda = -P - G_TF_{TT}^{-1}d_T, \tag{27b}$$

$$\gamma + (T_T^TF_T^{-1}F_{TN} - T_N^T)\, r_N + T_T^TF_T^{-1}G_T^T\mathring{u}^+ - T_T^TF_T^{-1}G_T^T\mathring{u}^-$$
$$- T_T^TF_{TT}^{-1}T_T\lambda = -T_T^TF_{TT}^{-1}d_T. \tag{28}$$

Finally a Linear Complementarity Problem (LCP) is obtained from equations (26)—(28) which can be written in the following form:

$$\left.\begin{aligned} &w - Mz = b, \\ &w \geqq 0, \quad z \geqq 0, \quad w^Tz = 0, \end{aligned}\right\} \tag{29}$$

where

$$w = \begin{bmatrix} y_N \\ v^+ \\ v^- \\ \gamma \end{bmatrix}, \quad z = \begin{bmatrix} r_N \\ \mathring{u}^+ \\ \mathring{u}^- \\ \lambda \end{bmatrix}, \quad b = \begin{bmatrix} d_N - F_{NT}F_{TT}^{-1}d_T \\ P + G_TF_{TT}^{-1}d_T \\ -P - G_TF_{TT}^{-1}d_T \\ -T_T^TF_{TT}^{-1}d_T \end{bmatrix},$$

$$M = \begin{bmatrix} (F_{NN} - F_{NT}F_{TT}^{-1}F_{TN}) & (G_N^T - F_{NT}F_{TT}^{-1}G_T^T) & -(G_N^T - F_{NT}F_{TT}^{-1}G_T^T) & F_{NT}F_{TT}^{-1}T_T \\ -(G_N - G_TF_{TT}^{-1}F_{TN}) & G_TF_{TT}^{-1}G_T^T & -G_TF_{TT}^{-1}G_T^T & -G_TF_{TT}^{-1}T_T \\ (G_N - G_TF_{TT}^{-1}F_{TN}) & -G_TF_{TT}^{-1}G_T^T & G_TF_{TT}^{-1}G_T^T & G_TF_{TT}^{-1}T_T \\ -(T_T^TF_{TT}^{-1}F_{TN} - T_N^T) & -T_T^TF_{TT}^{-1}G_T^T & T_T^TF_{TT}^{-1}G_T^T & T_T^TF_{TT}^{-1}T_T \end{bmatrix}.$$

4. On the existence of the solution

In this section we derive necessary and sufficient conditions which guarantee that the LCP (29) is solvable. They are based solely on the theory of FICHERA [17] for nonsymmetric variational inequalities. Here M is nonsymmetric. We note first that (see [6] p. 352) the LCP is equivalent to the following variational inequality:

$$(Mz + b)^T (z^* - z) \geqq 0, \qquad z \geqq 0, \qquad z^* \geqq 0. \tag{30}$$

Let us denote by L the intersection of the null space of $\boldsymbol{M}$ and the null space of $\boldsymbol{b}$, and L_1 a subspace of the null space of $\boldsymbol{M}$ such that

$$\mathfrak{R} = L \oplus L_1 , \tag{31}$$

$$\mathfrak{R} = \{\varrho \mid M\varrho = 0\} \tag{32}$$

and moreover let

$$\mathfrak{R}_1 = \mathfrak{R} \cap \{\varrho \mid \varrho \geqq 0\} . \tag{33}$$

Since $\boldsymbol{M}$ is nonnegative, the null space of $\boldsymbol{M}$ coincides with the null space of $(\boldsymbol{M} + \boldsymbol{M}^{\mathrm{T}})$.

Let

$$\mathfrak{R}' = \{\varrho \mid (\boldsymbol{M} + \boldsymbol{M}^{\mathrm{T}})\,\varrho = \boldsymbol{0}\} \tag{34}$$

and

$$\mathfrak{R}'_1 = \mathfrak{R}' \cap \{\varrho \mid \varrho \geqq 0\} . \tag{35}$$

Let us denote by $\mathfrak{R}^c$ the orthogonal complement of $\mathfrak{R}$ with respect to $\mathfrak{R}'$ because $\mathfrak{R} \subset \mathfrak{R}'$.

The following propositions give necessary and sufficient conditions for the existence of a solution of problem (29) or (30) and are special cases of the general proposition by FICHERA [17].

Proposition 1: *If the* LCP (29) *or the equivalent variational inequality* (30) *has a solution z for any ϱ such that $\varrho \in \mathfrak{R}_1$ and $\boldsymbol{b}^{\mathrm{T}}\varrho \neq \boldsymbol{0}$ one must have*

$$\boldsymbol{b}^{\mathrm{T}}\varrho > \boldsymbol{0} . \tag{36}$$

Proof: If (29) or (30) has a solution $z \geqq \boldsymbol{0}$, let us put in (30) $z^* = z_0 + \lambda\varrho \geqq \boldsymbol{0}$, $\lambda > \boldsymbol{0}$. Then we have

$$(\boldsymbol{M}z + \boldsymbol{b})^{\mathrm{T}}\,(z_0 + \lambda\varrho - z) \geqq 0 \quad \text{for all} \quad z_0 \geqq \boldsymbol{0}$$

or equivalently

$$\frac{(\boldsymbol{M}z + \boldsymbol{b})^{\mathrm{T}}\,(z_0 - z)}{\lambda} + \boldsymbol{b}^{\mathrm{T}}\varrho \geqq 0 .$$

As $\lambda \to \infty$ we have that

$$\boldsymbol{b}^{\mathrm{T}}\varrho \geqq \boldsymbol{0} . \tag{37}$$

Due to the assumption $\boldsymbol{b}^{\mathrm{T}}\varrho \neq \boldsymbol{0}$ we obtain (36). q.e.d.

Proposition 2: *Suppose that*

i) $\boldsymbol{b}^{\mathrm{T}}\varrho > 0$ *for all* $\varrho \in \mathfrak{R}_1$.

ii) *For every ϱ such that $Q_0\varrho \neq 0$, where Q_0 is the orthogonal projector of R^n onto $\mathfrak{R}^c$, and for $\varrho \in \mathfrak{R}'_1$ we have that there exists $z_\varrho \geqq \boldsymbol{0}$ such that*

$$-\boldsymbol{b}^{\mathrm{T}}\varrho + z_\varrho^{\mathrm{T}}\boldsymbol{M}\varrho < \boldsymbol{0} . \tag{38}$$

Then (30) *admits at least one solution.*

For the proof of this proposition we refer to ([17] proposition 2.1).

Proposition 3: *Suppose that for some ϱ satisfying the conditions $\varrho \in \mathfrak{R}'_1$, we have*

$$-\boldsymbol{b}^{\mathrm{T}}\varrho + z^{\mathrm{T}}\boldsymbol{M}\varrho > \boldsymbol{0} \quad \textit{for all} \quad z \geqq \boldsymbol{0} , \tag{39}$$

then the LCP (29) *has no solution.*

Proof: Let us suppose that the problem (29) or (30) has a solution $z \geqq \boldsymbol{0}$. We consider, for every $\lambda > \boldsymbol{0}$, the vector $z_0 + \lambda\varrho \geqq \boldsymbol{0}$ where $\varrho \in \mathfrak{R}'_1$ and let us put in (30) $z^* = z_0 + \lambda\varrho$. Then we have

$$((\boldsymbol{M} + \boldsymbol{M}^{\mathrm{T}} - \boldsymbol{M}^{\mathrm{T}})\,z + \boldsymbol{b})^{\mathrm{T}}\,(\boldsymbol{z}_0 + \lambda\varrho - z) \geqq \boldsymbol{0} ,$$
$$((\boldsymbol{M} + \boldsymbol{M}^{\mathrm{T}})\,z)^{\mathrm{T}}\,(z_0 + \lambda\varrho - z) \geqq (\boldsymbol{M}^{\mathrm{T}}z - \boldsymbol{b})^{\mathrm{T}}\,(z_0 + \lambda\varrho - z)$$

or equivalently

$$z^{\mathrm{T}}(\boldsymbol{M} + \boldsymbol{M}^{\mathrm{T}})\,(z_0 - z) + z^{\mathrm{T}}(\boldsymbol{M}\lambda\varrho + \boldsymbol{M}^{\mathrm{T}}\lambda\varrho) \geqq (\boldsymbol{M}^{\mathrm{T}}z - \boldsymbol{b})^{\mathrm{T}}\,\lambda\varrho + (\boldsymbol{M}^{\mathrm{T}}z - \boldsymbol{b})^{\mathrm{T}}\,(z_0 - z) .$$

Since $\varrho \in \mathfrak{R}'_1$, we have $\boldsymbol{M}\lambda\varrho + \boldsymbol{M}^{\mathrm{T}}\lambda\varrho = \boldsymbol{0}$ accordingly

$$z^{\mathrm{T}}(\boldsymbol{M} + \boldsymbol{M}^{\mathrm{T}})(z_0 - z) - (\boldsymbol{M}^{\mathrm{T}}z - \boldsymbol{b})^{\mathrm{T}}(z_0 - z) \geqq \lambda(\boldsymbol{M}^{\mathrm{T}}z - \boldsymbol{b})^{\mathrm{T}}\varrho,$$

$$\frac{z^{\mathrm{T}}(\boldsymbol{M} + \boldsymbol{M}^{\mathrm{T}})(z_0 - z) - (\boldsymbol{M}^{\mathrm{T}}z - \boldsymbol{b})^{\mathrm{T}}(z_0 - z)}{\lambda} \geqq (\boldsymbol{M}^{\mathrm{T}}z - \boldsymbol{b})^{\mathrm{T}}\varrho.$$

Thus as $\lambda \to \infty$ we obtain that

$$(\boldsymbol{M}^{\mathrm{T}}z - \boldsymbol{b})^{\mathrm{T}}\varrho \leqq 0$$

which contradicts (39).

Proposition 4: *If* $z^{\mathrm{T}}\boldsymbol{M}z$ *is positive definite on* R^n *i.e.:*

$$z^{\mathrm{T}}\boldsymbol{M}z > 0 \quad \textit{for all} \quad z \in R^n \quad \textit{and} \quad z \neq \boldsymbol{0}, \tag{40}$$

the LCP (29) *or equivalently* (31) *has only one solution for any given* $\boldsymbol{b}$.

Proof: For the existence of the solution (cf. [17] pp. 398). To show the uniqueness of the solution we assume that the problem has two solutions z_1 and z_2 and setting in (30) $z^* = z_1$, $z = z_2$ and then $z^* = z_2$, $z = z_1$ and adding that

$$(z_1 - z_2)^{\mathrm{T}}\boldsymbol{M}(z_1 - z_2) \leqq 0,$$

we obtain due to to (40) that $z_1 = z_2$ i.e. the uniqueness of the solution.

5. Numerical examples

In this section, numerical examples of grasping with frictional grippers are presented. For all the examples the frictionless finger contact is modeled to have a linear spring which is oriented normal on the object surface. A hard-finger contact has three linear springs, one normal and the two others on the tangential directions. A soft-finger contact has, in addition, an angular spring with axis oriented along the normal on the object boundary and through the point of contact [5]. The flexibility of the springs are, for the normal $F_{\mathrm{NN}i} = 1.5\mathrm{E} - 6$, the tangential $F_{\mathrm{TT}1i} = F_{\mathrm{TT}2i} = 6.0\mathrm{E} - 6$ and for the angular $F_{\mathrm{TT}mi} = 1.0\mathrm{E} - 5$. The initial openings d_{N} and d_{T} are assumed to be equal to zero. In the approximation of the friction cone we use eight surfaces.

As a first example let us consider the three-dimensional object which grasped with three fingers as shown in Fig. 3. For this example we consider three cases:

i) Case 1: In this case we assume that finger № 1 is a frictionless and the two others are hard-fingers. Finger № 1 exerts a normal force $r_{\mathrm{N}1} = 1$ on the object. For this force, and with friction coefficient $\mu = 0.3$, the normal and the tangential reactions of fingers 2 and 3, the corresponding slipping values and the rigid body displacements and rotations are given in Table 1.

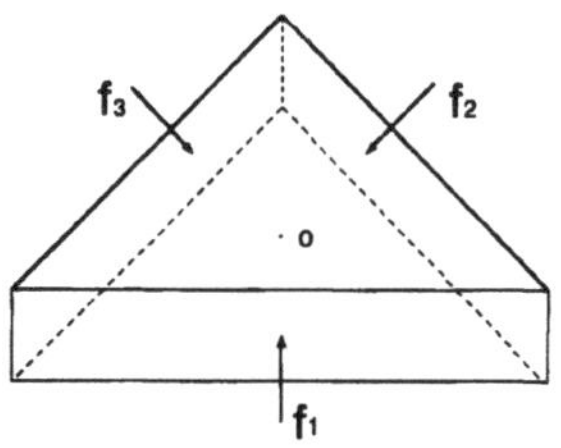

Fig. 3. Configuration of example 1

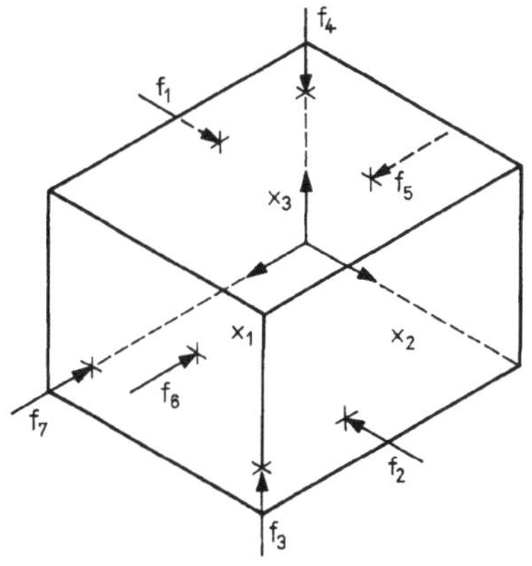

Fig. 4. Configuration of example 2

ii) Case 2: Here all the fingers are assumed to be hard-fingers. Finger 1 exerts on the object the forces $r_{\mathrm{N}1} = 10.$, $r_{\mathrm{T}11} = 1.$ and $r_{\mathrm{T}21} = 1.$. The friction coefficient is $\mu = 0.3$. In Table 2 the reactions and the corresponding slipping values of fingers 2 and 3, and the rigid body displacements and rotations are given.

Table 1. Example of Fig. 3, case 1

fingers	contact reactions	gaps	tangential components	frictional forces	slipping values	rigid body motion components	rigid body motion values
2	5.660E−01	0.000E+00	1	−1.410E−01	0.000E+00	$\mathring{u}_1$	8.480E−07
			2	−0.000E+00	0.000E+00	$\mathring{u}_2$	8.480E−07
3	5.660E−01	0.000E+00	1	−1.410E−01	0.000E+00	$\mathring{u}_3$	0.000E+00
			2	−0.000E+00	0.000E+00	$\mathring{\varphi}_1$	0.000E+00
						$\mathring{\varphi}_2$	0.000E+00
						$\mathring{\varphi}_3$	0.000E+00

Table 2. Example of Fig. 3, case 2

fingers	contact reactions	gaps	tangential components	friction forces	slipping values	rigid body motion components	rigid body motion values
2	5.660E+00	0.000E+00	1	−1.410E−01	0.000E+00	$\mathring{u}_1$	8.480E−06
			2	5.000E−01	0.000E+00	$\mathring{u}_2$	8.480E−06
3	5.660E+00	0.000E+00	1	−1.410E−01	0.000E+00	$\mathring{u}_3$	−3.000E−06
			2	5.000E−01	0.000E+00	$\mathring{\varphi}_1$	0.000E+00
						$\mathring{\varphi}_2$	0.000E+00
						$\mathring{\varphi}_3$	0.000E+00

Table 3. Example of Fig. 3, case 3

fingers	contact reactions	gaps	tangential components	friction forces	slipping values	rigid body motion components	rigid body motion values
2	5.656E+00	0.000E+00	1	6.989E−01	0.000E+00	$\mathring{u}_1$	8.484E−06
			2	1.201E+00	0.000E+00	$\mathring{u}_2$	1.165E−05
			3	−9.883E−01	0.000E+00	$\mathring{u}_3$	−3.122E−05
3	7.769E+00	0.000E+00	1	−1.414E+00	0.000E+00	$\mathring{\varphi}_1$	5.929E−06
			2	6.989E−01	0.000E+00	$\mathring{\varphi}_2$	−1.698E−05
			3	1.698E+00	0.000E+00	$\mathring{\varphi}_3$	−1.120E−05

iii) Case 3: Now we assume that the fingers 2 and 3 are soft-fingers and finger 1 is a frictionless one. Finger 1 exerts a normal force $r_{N1} = 10.$, also an external force (e.g. the object weight) $P_3 = 1.9$ is applied. The friction coefficient $\mu = 0.3$ and the torsional friction coefficient $\mu_m = 0.5$. Here we must note that in cases 1 and 2 when the external force P_3 is applied the object can not be grasped. The reactions and the slipping values of fingers 2 and 3 are given in Table 3.

To give another example we consider a rectangular prism which is grasped by seven potential hard-fingers as shown in Fig. 4. Let us assume that the following external forces are applied on the object ($P_1 = 10.$, $P_2 = 0.$, $P_3 = 30.$, $M_1 = 30.$, $M_2 = 0.$, $M_3 = 45.$). The contact reactions, the friction forces and the corresponding slipping values, also the rigid body displacements and rotations are given in Table 4 for friction coefficient $\mu = 0.3$, and in Table 5 for friction coefficient $\mu = 0.5$.

Table 4. Example of Fig. 4, with friction coefficient $\mu = 0.3$

fingers	contact reactions	gaps	tangential components	friction forces	slipping values	rigid body motion components	rigid body motion values
1	1.832E+00	0.000E+00	1	−2.103E−01	1.835E−04	u_1	−1.793E−04
			2	5.508E−01	−1.908E−04		
2	0.000E+00	2.895E−04	1	0.000E+00	2.967E−04	u_2	7.291E−05
			2	0.000E+00	8.207E−05		
3	1.190E+01	0.000E+00	1	−3.299E+00	4.118E−04	u_3	−2.608E−05
			2	1.366E+00	−1.681E−04		
4	2.415E+01	0.000E+00	1	−6.693E+00	1.872E−04	φ_1	−5.599E−05
			2	−2.772E+00	6.909E−06		
5	4.130E+01	0.000E+00	1	9.439E−01	0.000E+00	φ_2	−5.073E−05
			2	1.032E+01	0.000E+00		
6	7.676E+00	0.000E+00	1	−8.813E−01	4.674E−05	φ_3	−9.268E−05
			2	0.2127E+01	−2.378E−04		
7	0.000E+00	5.933E−05	1	0.000E+00	−4.030E−05		
			2	0.000E+00	−1.976E−05		

Table 5. Example of Fig. 4, with friction coefficient $\mu = 0.5$

fingers	contact reactions	gaps	tangential components	friction forces	slipping values	rigid body motion components	rigid body motion values
1	1.765E−02	0.000E+00	1	−3.378E−03	1.285E−04	u_1	−1.122E−04
			2	8.156E−03	−1.506E−04		
2	0.000E+00	2.066E−04	1	0.000E+00	−2.028E−04	u_2	5.432E−05
			2	0.000E+00	7.640E−01		
3	1.006E+01	0.000E+00	1	4.648E+00	−2.802E−04	u_3	−3.927E−05
			2	−1.925E+00	1.066E−04		
4	1.845E+01	0.000E+00	1	−8.526E+00	9.725E−05	φ_1	−3.714E−05
			2	4.293E−02	0.000E+00		
5	3.615E+01	0.000E+00	1	−2.389E+00	0.000E+00	φ_2	−3.955E−05
			2	6.460E+00	0.000E+00		
6	4.196E+00	0.000E+00	1	8.030E−01	−4.204E−05	φ_3	−6.378E−05
			2	−1.938E+00	1.565E−04		
7	0.000E+00	5.544E−05	1	0.000E+00	1.660E−05		
			2	0.000E+00	9.455E−06		

References

1 CUTKOSKY, M. R.: On grasp choice, grasp models, and the design of hands for manufactoring tasks. IEEE Trans. Robotics and Automation **5** (1989) 3, 269−279.

2 KERR, J.; ROTH, B.: Analysis of multifingered hands. Internat. J. of Robotics Res. **4** (1986) 4, 3−17.

3 MASON, M. T.; SALISBURY, J. K.: Robot hands and the mechanics of manipulation. MIT Press, Cambridge MA 1985.

4 KUMAR, V.; WALDRON, K. J.: Suboptimal algorithms for force distribution in multifingered grippers. IEEE Trans. Robotics and Automation **5** (1989) 4, 491−498.

5 NGUYEN, V-D.: Constructing stable grasps. Internat J. of Robotics Res. **8** (1989) 1, 26−36.

6 PANAGIOTOPOULOS, P. D.: Inequality problems in mechanics and applications. Convex and nonconvex energy functions. Birkhäuser Verlag, Basel-Boston 1985 (Russian translation, Moscow: MIR Publ. 1989).

7 PANAGIOTOPOULOS, P. D.: A nonlinear programming approach to the unilateral contact-, and friction-boundary value problem in the theory of elasticity. Ingenieur-Arch. **44** (1975), 421−432.

8 ODEN, J. T.; MARTINS, J. A. C.: Models and computational methods for dynamics friction phenomena. Computer Methods in Appl. Mech. and Engng. **52** (1985), 527−634.

9 ODEN, J. T.; PIRES, E. B.: Nonlocal and nonlinear friction laws and variational principles for contact problems in elasticity. J. Appl. Mech. **50** (1983) 1, 67−76.

10 KLARBRING, A.; BJÖRKMAN, G.: A mathematical programming approach to contact problems with friction and varying contact surface. Computers and Structures **30** (1988) 5, 1185−1198.

11 BISBOS, C. D.: A discrete nonlinear boundary variational inequality for 3-D frictional contact problems. Internal report 1542, Institute of Steel Structures, Aristotle University, Thessaloniki 1990.

12 MITSOPOULOU, E. N.; DOUDOUMIS, I. N.: A contribution to the analysis of unilateral contact problems with friction. Solid Mechanics Archives **12** (1987) 2, 165−186.

13 MARKENSCOFF, X.; NI, L.; PAPADIMITRIOU, CH.: The geometry of grasping. Internat. J. of Robotics Res. **9** (1990) 1, 61−74.

14 MURTY, K. G.: Linear complementarity. Linear and nonlinear programming. Heldermann Verlag, Berlin 1988.

15 LEMKE, C. A.: Some pivot schemes for the Linear Complementarity Problem. In: BALINSKI, M. L.; COTTLE, R. W. (eds.): Mathematical programming studies **7**, 1978, pp. 15−35.

16 HOLMBERG, G.: A solution scheme for three-dimensional multi-body contact problems with friction using mathematical programming. Thesis No. 195, Dept. of Mechanical Engineering, Linköping Institute of Technology, Sweden 1989.

17 FICHERA, G.: Boundary value problems in elasticity with unilateral constraints. In. FLÜGGE, S. (eds.): Encyclopedia of Physics, Vol. VI a/2, Springer Verlag, Berlin 1972.

Received June 11, 1990

Addresses: A. M. AL-FAHED, Dipl. Ing. Research Assistant, School of Technology, Aristotle University, GR-54006 Thessaloniki, Greece,
G. E. STAVROULAKIS, Dipl. Ing. Research Assistant, School of Technology, Aristotle University, GR-54006 Thessaloniki, Greece, and
P. D. PANAGIOTOPOULOS, Ord. Professor Dr. Ing., School of Technology, Aristotle University, GR-54006 Thessaloniki, Greece and Honorarprofessor, Faculty of Mathematics and Physics, R.W.T.H., W-51000 Aachen, F. R. Germany

BOOK REVIEWS

Parker, T. S.; Chua, L. O., Practical Numerical Algorithms for Chaotic Systems. Berlin etc., Springer-Verlag 1989. XIV, 348 S., 152 Abb., DM 98.–. ISBN 3-540-96689-7

Die Autoren, zwei der hervorragendsten Vertreter der nichtlinearen Dynamik, stellen sich mit dem vorliegenden Buch ein hohes Ziel: Ohne vorausgesetzte Kenntnisse über nichtlineare Systeme das grundlegende Verhalten chaotischer Systeme auf intuitive, doch mathematisch gesicherte Weise zu erklären sowie zuverlässige Algorithmen zu ihrer Simulation und Charakterisierung zu entwikkeln. Sie gehen von der Tatsache aus, daß heutzutage jeder mit einem PC nichtlineare Systeme simulieren kann, andererseits eine Simulation kaum zu theoretischen Ergebnissen führen kann; selbst wenn sie absolut genau wäre, ergäbe sie lediglich *eine* Lösung *eines* Systems mit *einem* Satz von Anfangsbedingungen und Parameterwerten. In den meisten Kapiteln des Buches schließen sich an einen theoretsichen Teil entsprechende in einem Pseudo-Code beschriebene Algorithmen an. Dazu wird ein in C geschriebenes unter UNIX und PC-DOS laufendes Softwarepaket INSITE (Interactive Nonlinear System Investigative Toolkit for Everyone) vorgestellt, eine Sammlung interaktiver grafisch orientierter Programme für nichtlineare und insbesondere chaotische Systeme, welches gesondert bezogen werden kann; ein genauer Vergleich von C und FORTRAN und deren gemeinsame Nutzung werden im einzelnen diskutiert.

In einem einleitenden Kapitel wie auch in Anhängen werden die Grundtatsachen dynamischer Systeme in ausgewogener Anschaulichkeit und Präzision, mit Hinweisen auf besonders geeignete weiterführende Literatur, behandelt. Es folgt die Bestimmung von Poincaré-Abbildungen, klar unterschieden für nichtautonome und autonome Systeme, mit Erläuterung und Vergleichen verschiedener Algorithmen (beispielsweise nach Newton-Raphson und nach Henon). Im Rahmen eines Stabilitätskapitels (mit dem Vorschlag, zwischen „unstable“ als asymptotisch stabil bei rückläufiger Zeit und „non-stable“ als auch rückläufig „non-stable“ zu unterscheiden) wird vor allem auf die Ljapunovexponenten und einen Algorithmus zu ihrer Berechnung eingegangen. Bei der Behandlung von Integrationsalgorithmen, Integrationsfehlern und steifen Gleichungen werden auch praktische Hinweise zu Integrationsalgorithmen und -routinen gegeben und die besonderen Schwierigkeiten bei der Integration chaotischer Systeme erklärt. Algorithmen werden auch für die Lokalisierung von Gleichgewichtspunkten, Fixpunkten, geschlossenen Orbits und periodischen Lösungen gegeben. Auch die Betrachtung stabiler und instabiler Mannigfaltigkeiten und des Satzes von Šilnikov werden durch entsprechende Algorithmen ergänzt. Ausführlich wird die Dimension seltsamer Attraktoren behandelt. Für verschiedene Dimensionsbegriffe werden vergleichende allgemeine Ableitungen und Berechnungsalgorithmen abgegeben. Auch für Verzweigungsdiagramme werden neben der Theorie Algorithmen dargestellt und zahlreiche praktische Hinweise gegeben. Das gleiche gilt für Phasenportraits und Anziehungsbereiche, die im letzten Kapitel behandelt werden.

Das Buch erfüllt voll das gesteckte Ziel, natürlich können aus Umfangsgründen nur wenige Ergebnisse über die zahlreichen bisher untersuchten konkreten Klassen chaotischer Systeme aufgenommen werden. Wer immer chaotische Systeme numerisch zu untersuchen hat, sollte die in dem Buch dargestellten umfangreichen Erfahrungen nutzen.

Berlin G. Schmidt

de Silva, C. W.; MacFarlane, A. G. J., Knowledge-Based Control with Application to Robots. Berlin etc., Springer-Verlag 1989. X, 196 pp., 41 figs., DM 56.–. ISBN 3-540-51143-1 (Lecture Notes in Control and Information Sciences 123)

Das vorliegende Buch stellt eine Integration von „hard“-algorithmischer und „soft“-wissensbasierter Regelung eines Roboterarms vor. Die hierarchische Anordnung der Steuerung geht aus von einem Hochgeschwindigkeitsregler auf dem untersten Niveau, einem intelligenten Beobachter auf dem mittleren sowie einem „Tuner“ auf dem oberen Niveau. Für den „Tuner“ wird ein Fuzzy-Regler eingesetzt, der die Parameter eines PID-Reglers entsprechend den „Vorschlägen“ eines intelligenten Beobachters über Entscheidungstabellen adaptiert.

Zunächst wird auf die allgemeinen kinematischen und dynamischen Zusammenhänge bei einem Manipulator eingegangen, wobei ein rekursives Verfahren zur Berechnung der Roboterdynamik favorisiert wird. Die im Anschluß folgende Einführung in die Fuzzy-Logik gibt Einblick in die Grundbegriffe der Theorie der Fuzzy-Sets und geht dann auf Probleme der Fuzzyzustandsgleichungen ein. Weiterhin wird besprochen, auf welche Weise die Zugehörigkeitsfunktionen ermittelt bzw. geschätzt werden, wobei hierzu fünf Methoden angegeben werden. Nach dieser Einführung wird eingehend die Steuerungsstruktur des Roboters beschrieben. Der Roboter besitzt eine sehr effektive Servoregelung auf dem untersten Niveau, einen Satz intelligenter Preprozessoren (Servo-Experten) für jede Roboterachse auf dem mittleren Niveau und ein „Fuzzy-Tuner“ auf dem obersten Niveau. Auf diese Weise werden die Vorteile von sogenannten „hard“-Algorithmen und wissensbasierten „soft“-Algorithmen vereinigt. Im folgenden werden die einzelnen Niveaus genauer vorgestellt:

Das Servo-Niveau besteht aus einem PID-Regler und einem nichtlinearen Kompensationsterm, bei dem nach dem Newton-Euler-Verfahren die nichtlineare Roboterdynamik mit einbezogen wird. Als Voraussetzung hierzu wird die Messung der Beschleunigung und der Reaktionskräfte in den Gelenken eingegeben. Das Servo-Experten-Niveau besteht aus je einem Servo-Experten für jeweils eine Achse. Es werden fünf Merkmale zur Beurteilung des Prozesses herangezogen: 1. Genauigkeit, 2. Oszillationen, 3. Konvergenzgeschwindigkeit des Fehlers, 4. Trend (Divergenz), 5. Statischer off-set.

Hierzu werden von dem Servo-Experten Qualitätseinschätzungen getroffen. Diese Einschätzungen erfolgen über einen längeren Zeithorizont, der nach Bedarf verändert werden kann. Die Einschätzung erfolgt durch Abarbeitung von Regeln wie IF E > EMAX THEN OKAY. Das Fuzzy-Regelungs-Niveau trifft auf der Basis der von den Servo-Experten vorgelegten Qualitätseinschätzungen mit Hilfe von unscharfen IF-THEN-Regeln Entscheidungen zur Veränderung der PID-Reglerparameter eines Gelenkes. Dieses Niveau ist ausbaufähig, so daß auch Lernstrukturen denkbar sind. Das zeitintensive on-line „Rule-matching“ wird abgelöst durch eine off-line-Vorausberechnung von Entscheidungstabellen, die on-line abgearbeitet werden. Voraussetzung hierzu sind diskret abspeicherbare Zugehörigkeitsfunktionen. Die Ausgangs-fuzzy-vektoren werden durch die Bildung des „gewichteten Mittels“ in skalare Ausgänge für die PID-Parameterkorrekturen umgewandelt. Das Verfahren wurde simulativ am Beispiel eines Roboters mit zwei rotatorischen Freiheitsgraden auf einer SUN – 3/50 – Workstation getestet. Programmiert wurde in der Sprache C unter UNIX. Die Servo-Experten nutzen das kommerziell verfügbare System MUSE AI Toolkit. Regeln werden in Poptalk geschrieben. Gegenüber einer konventionellen Regelung zeigt die neue Regelung eine höhere Qualität bei größerer Robustheit hinsichtlich Systemparameterveränderungen. Der Fuzzy-Regler identifiziert die auftretenden Trends der auftretenden Signale, Oszillationen usw. Servoregler und Fuzzy-Tuner arbeiten mit Taktzeiten, die sich um Größenordnungen unterscheiden.

Das Buch ist für einen auf dem Gebiet der Robotertechnik tätigen Fachmann flüssig und einleuchtend geschrieben. Die eingehende Behandlung der Abschnitte zur Fuzzy-Logik sowie zur Implementierung ermöglicht ein Nachvollziehen der einzelnen Realisierungsschritte. Leider liegen praktische Resultate an einem realen Roboter nicht vor, so daß ausschließlich Simulationen den Prüfstein für das vorgestellte System bilden.

Berlin R. Palm

ZAMM · Z. angew. Math. Mech. 71 (1991) 7/8, 267–276 Akademie Verlag

DRESSLER, R.; ISMAR, H.; RIPPLINGER, W.

Untersuchungen zur werkstoffmechanischen Stabilität metallischer Faserverbundwerkstoffe mit geschädigter Struktur bei zyklischer thermomechanischer Belastung

Herrn Professor Dr.-Ing. OSKAR MAHRENHOLTZ zum 60. Geburtstag gewidmet

Das Verhalten eines faserverstärkten Bauteils mit metallischer Matrix wird unter zyklischer thermomechanischer Belastung simuliert. Die metallische Matrix des Verbundwerkstoffes wird durch ein weiterentwickeltes Zweiflächenverfestigungsmodell auf der Grundlage des Übergangsfließpotentials beschrieben, um die typischen nicht-linearen Effekte zyklischer Plastizität und deren Temperaturabhängigkeit zu berücksichtigen. Es wird eine spröde, intermetallische Schicht zwischen Faser und Matrix angenommen, die sich linear-elastisch verhält. Ohne ein neues FE-Netz erzeugen zu müssen, können wir numerisch das Fortschreiten eines anfänglich existierenden Risses innerhalb der Zwischenschicht verfolgen. Auf der Grundlage dieses Werkstoffmodells untersuchen wir die werkstoffmechanische Stabilität an dem ausgewählten Beispiel eines faserverstärkten Rohres.

The behaviour of a metal-matrix-composite under cyclic thermo-mechanical loading is carefully simulated. The metal-matrix of the composite is described by an advanced two- surface workhardening model on the basis of the transition yield potential, in order to take into account for the typical nonlinear effects of cyclic plasticity and temperature-dependence of yield stress. A brittle intermetallic layer between fibre and matrix is assumed to behave linear-elasticly. Without the disadvantage of having to generate a new FE-grid, we can watch numerically the propagation of an existing crack within the interface layer. On the basis of these material models we are interested in the material stability of the selected example of a fibre reinforced tube.

Внимательно симмулируется поведение металло-матричного составного под циклическом термомеханическом напряжением. Металлическая матрица составного описывается при помощи продвигнотой двухповерхностной модели упрочения на основе переходного потенциала текучести чтобы принимать в расчет тиличные нелинейные эффекты циклической упругости и температурной зависимости предела текучести. Предполагается, что хупкий интерметаллический слой между волокном и матруцей поступает линейно-упрутим. Без невыгоды надобности образования новой конечно-элементной сетки можно смотреть численном образом на рагпространение существовающей трещины в промежуточном слое. На основе этой модели материала мы в интересном материальной устойчивости отобранного примера усиливанной волокном трубки.

MSC (1980): 73K20

1. Einleitung

Es ist absehbar, daß in technischen Strukturen bei großen thermomechanischen Belastungen zunehmend metallische Faserverbundwerkstoffe Eingang finden. Erfolgt die Beanspruchung zyklisch, so ist für eine Lebensdauerprognose die werkstoffmechanische Stabilität nach vorangegangener inelastischer Beanspruchung von besonderer Bedeutung. Dabei stellt sich bei realen Werkstoffen zusätzlich die Frage, inwieweit eine Schädigung im Bereich der Phasengrenzen durch Risse diese Stabilität beeinflussen wird.

Dazu vorab eine kurze Beschreibung des Aufbaus gängiger Verbundstrukturen. Faserverbundwerkstoffe mit metallischer Matrix bestehen in der Regel aus drei Phasen, den Fasern (F), den Zwischenschichten (Reaktionsschichten) (Z) und der Matrix (M). Bekannte Vertreter dieser Stoffe (F/Z/M) sind beispielsweise C/Ni/AlMgSi, Al_2O_3/–/AlMgSi oder auch X2NiCoMo/Al_5Fe_2/AlMgSi. Bei den Kohlenstoffasern (C) in einer Aluminiummatrix wird als Benetzungsvermittler Nickel eingesetzt. Aluminiumoxidfasern (Al_2O_3) nehmen wegen ihrer chemischen Verträglichkeit zum Matrixwerkstoff Aluminium eine besondere Stellung ein. Hier wird der Benetzungsvermittler SiO_2 im Fertigungsprozeß von der schmelzflüssigen Aluminiumlegierung wieder gelöst, so daß ein zweiphasiger Verbundwerkstoff vorliegt. Bei hochfesten Marageing-Stahlfäden, die ebenfalls in eine Aluminiummatrix eingesetzt werden, kommt es an den Phasengrenzen zur Bildung einer spröden (elastischen) Reaktionsschicht aus der intermetallischen Verbindung Al_5Fe_2. An diesen drei ausgewählten Beispielen kann man bereits erkennen, daß aus werkstoffphysikalischer Sicht den Phasengrenzen eine große Bedeutung zukommt.

In der späteren Fallstudie soll nun exemplarisch für eine Verbundstruktur aus Stahlfasern in einer Aluminiummatrix die werkstoffmechanische Stabilität bei zyklischer Beanspruchung numerisch untersucht werden. Die Struktur wird dabei mikromechanisch mit Hilfe der FEM betrachtet. Ein besonderer Schwerpunkt ist die inelastische Modellierung durch Einführung eines Übergangsfließpotentials in die Zweiflächenverfestigungstheorie. Risse in der spröden Zwischenschicht werden durch anisotropes Absenken der elastischen Eigenschaften im finiten Rißelement und durch die Einführung eines Bruchlastvektors in das FE-Gesamtgleichungssystem simuliert. Die thermomechanische Kopplung wird über die thermische Ausdehnung im Verschiebungsgleichungssystem einerseits und über die Zuführung plastisch dissipierter Arbeit ins Temperaturgleichungssystem andererseits vorgenommen.

2. Grundlagen

Inelastisches Stoffmodell

Die Modellierung von Faserverbundwerkstoffen erfolgt heute überwiegend nach zwei Methoden, der Makro- oder Kontinuumsmethode und der Mikromethode. In der ersten Methode werden die mechanischen Eigenschaften der einzelnen Phasen verschmiert. Damit läßt sich das anisotrope, makromechanische Verhalten der Struktur gut abschätzen. Aussagen über lokale Ereignisse sind aber nicht möglich. Bei der mikromechanischen Betrachtungsweise stehen dagegen die aktuelle Geometrie und die stoffmechanischen Eigenschaften der einzelnen isotropen Phasen im Vordergrund. Diese Vorgehensweise wurde von HASHIN [1, 2] begründet. Sie soll auch in der vorliegenden Studie Anwendung finden.

Die elastisch-plastische Stoffmodellierung der isotropen Werkstoffphasen wird nach PRANDTL-REUSS [3] unter Verwendung eines Übergangsfließpotentials [4] vorgenommen. Erreicht der Spannungsvektor im Spannungsraum die Fließfläche, dann ist nach diesem Modell das Inkrement der plastischen Formänderung dem Inkrement der elastischen Formänderung zu überlagern:

$$\mathrm{d}\varepsilon = D^{-1}\,\mathrm{d}\sigma + \mathrm{d}\lambda\,\frac{\partial F}{\partial \sigma}\,. \tag{2.1}$$

Der erste Teil der Gleichung (2.1) für das Gesamtformänderungsinkrement dε ist das linearelastische Hooksche Gesetz. Der zweite Teil enthält das Potentialgesetz nach v. Mises in Verallgemeinerung auf das Übergangsfließpotential.

Bei Berücksichtigung isotrop-kinematischer Verfestigung lautet unser Fließpotentialansatz [5]:

$$F = \sqrt{\frac{\alpha(\varepsilon_v^p)\,\{s_{kk}\}^2 + \frac{1}{2}\,s'_{ij}s'_{ij}}{\alpha(\varepsilon_v^p) + \frac{1}{3}}} - R(\varepsilon_v^p, T) = 0\,. \tag{2.2}$$

Gleichung (2.2) spannt im Hauptspannungsraum eine Fließfläche auf, deren Mitte als Folge der kinematischen Verfestigung um ζ verschoben ist. Mit den Translationskoordinaten ζ_{ij} und den Spannungskoordinaten σ_{ij} erhält man die Tensorkoordinaten $s_{ij} = \sigma_{ij} - \zeta_{ij}$ und daraus die Deviatorkoordinaten $s'_{ij} = s_{ij} - \frac{1}{3}\,s_{kk}\delta_{ij}$. Die Größe $R(\varepsilon_v^p, T)$ berücksichtigt den isotropen Verfestigungseinfluß in Abhängigkeit von der plastischen Vergleichsdehnung ε_v^p und der Temperatur T. $\alpha(\varepsilon_v^p)$ ist eine experimentell ermittelte Ansatzfunktion, die nachfolgend erläutert wird.

Bild 1 zeigt exemplarisch für einen metallischen Werkstoff (hier Stahl) die typischen Verläufe des Ansatzwertes α und der Fließspannung (Formänderungsfestigkeit) k_f bei zyklisch-schwellender Belastung und Raumtemperatur in Abhängigkeit von der plastischen Vergleichsdehnung ε_v^p. Derartige Kurven liegen für mehrere Aluminium- und Stahllegierungen bei verschiedenen Temperaturprofilen vor. Die Messungen wurden an isotropen Rohrproben in unterschiedlichen Quadranten des Hauptspannungsraumes vorgenommen und nach Gleichung (2.2) direkt „online" ausgewertet. Man sieht, daß bei beginnender Plastifizierung die Werte mit $\alpha > 0$ beginnen, um dann in einem sehr kleinen Formänderungsbereich

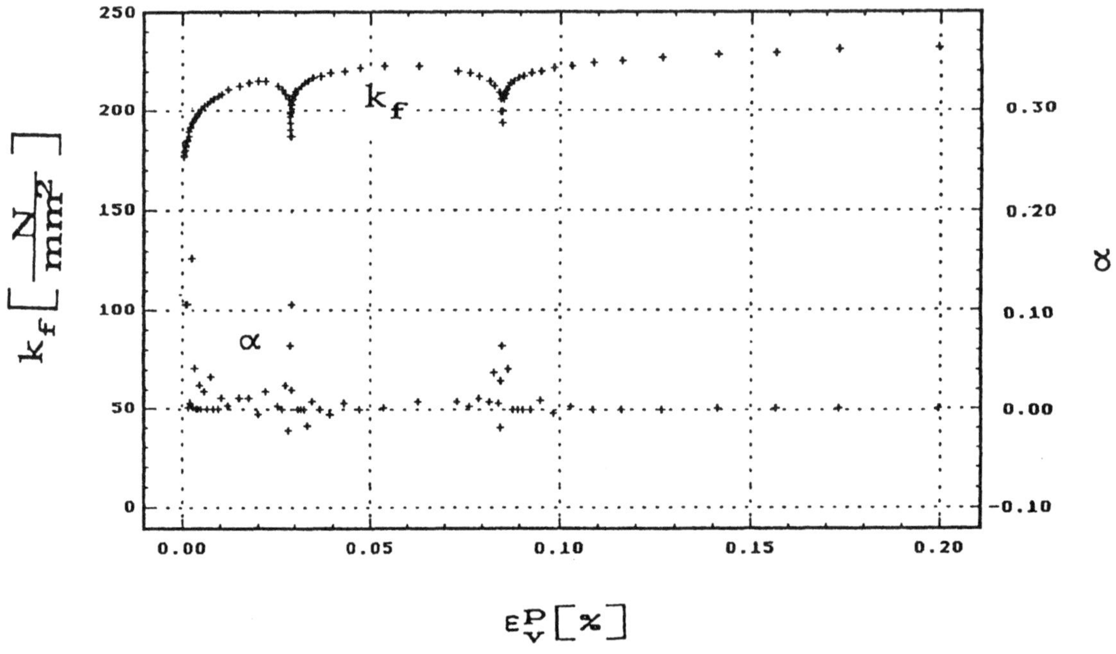

Bild 1. Typischer Verlauf der Fließspannung k_f und des Ansatzwertes α für metallische Werkstoffe (hier Stahl) in Abhängigkeit von der plastischen Vergleichsdehnung ε_v^p

auf $\alpha = 0$ abzufallen. Dies bedeutet, daß die Fließfläche im Übergangsbereich von der Form eines Rotationsellipsoiden in einen v. Mises-Zylinder übergeht. In fortlaufenden Entlastungs- und Belastungszyklen wiederholt sich dieser Vorgang. Bei realen Werkstoffen gehen somit zyklische Belastungen mit raschen Änderungen der Fließspannung und der Fließflächengeometrie einher. Andererseits sollte nicht verschwiegen werden, daß dieser Effekt oft von Störungen überlagert wird, besonders dann, wenn lokale elastische Potentiale oder Inhomogenitäten im Gefüge die Plastifizierung beeinflussen.

Bei der werkstoffmechanischen Beschreibung zyklischer, inelastischer Belastungsvorgänge kommt dem Verfestigungsmodell eine große Bedeutung zu. Die wichtigsten Effekte, die das Stoffmodell simulieren sollte, sind die zyklische Verfestigung, die Hysterese, die Bauschinger-Effekte und die zyklisch induzierte Spannungsrelaxation. Sie werden z. Z. am besten durch die Zweiflächenverfestigungstheorie erfaßt, die in ihrer mathematischen Beschreibung auf DAFALIAS und POPOV [6] zurückgeht und in der Folgezeit von verschiedenen Autoren weiterentwickelt wurde. Diese Theorie beschränkt sich aber z. Z. auf Fließflächen mit quadratischem Ansatz in den Spannungsdeviator-Koordinaten. Ihre Verallgemeinerung ist daher u. a. ein Schwerpunkt dieser Arbeit, da hier das Übergangsfließpotential in die Zweiflächentheorie implementiert werden soll. Damit liegen bei beginnender Plastifizierung zunächst zwei identische Rotationsellipsoidflächen vor, nämlich die aktuelle Fließfläche und die Grenzfläche. Diese Flächen entwickeln sich unterschiedlich mit fortschreitendem Fließen. Die Änderung der aktuellen Fließfläche erfolgt durch isotrop-kinematische Verfestigung, während die Grenzfläche überwiegend isotrop der Verfestigung einer monotonen Zugkurve gleicher plastischer Vergleichsdehnung folgt. Da sich die Grenzfläche stärker als die Fließfläche entwickelt, liegt letztere immer innerhalb der Grenzfläche.

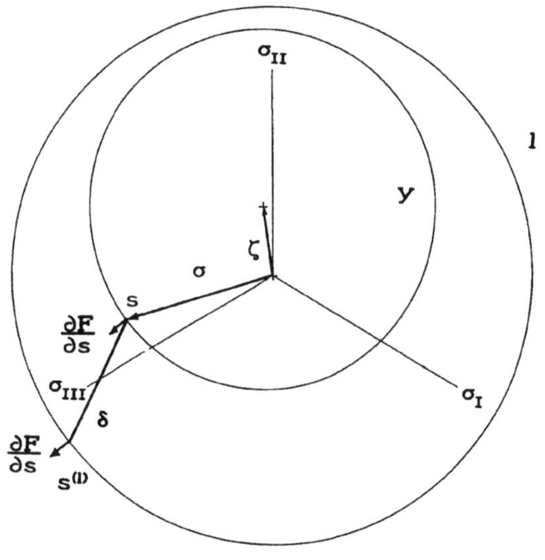

Bild 2. Aktuelle Fließfläche (y) und Grenzfläche (1) im Spannungsraum

Bild 2 zeigt in einer Momentaufnahme in Richtung σ_{ii} beide Flächen im Hauptspannungsraum. Der isotrope Verfestigungszuwachs der aktuellen Fließfläche (y) ist gegeben durch

$$dR^{(y)} = \left.\frac{\partial k_f}{\partial T}\right|_{\varepsilon_v^p = 0} dT + \xi^{(y)} \left[E^p(\delta)\, d\varepsilon_v^p - \left.\frac{\partial k_f}{\partial T}\right|_{\varepsilon_v^p = 0} dT \right]. \tag{2.3}$$

Er wird bestimmt durch den plastischen Modul $E^p(\delta)$, der in jedem Lastschritt neu zu berechnen ist als Funktion des Abstandes δ vom Spannungspunkt „s“ auf der aktuellen Fließfläche zum korrespondierenden Punkt „s^l“ auf der Grenzfläche. Die Punkte haben dabei auf beiden Flächen die gleiche Normale. Da $E^p(\delta) \geqq \frac{\partial k_f}{\partial \varepsilon_v^p}$, nimmt der plastische Modul nur Werte oberhalb der monotonen Steigung der einhüllenden Fließkurve an. ξ ist ein Wichtungsfaktor für den Anteil der isotropen Verfestigung.

Die kinematische Verfestigung folgt aus der Verschiebung ζ der Fließfläche. Die Zuwächse dieser Translation sind gegeben durch

$$d\zeta_{ij}^{(y)} = (1 - \xi^{(y)}) \left[E^p(\delta)\, d\varepsilon_v^p - \left.\frac{\partial k_f}{\partial T}\right|_{\varepsilon_v^p = 0} dT \right] \frac{\varkappa_{ij}}{\varkappa_{kl} \dfrac{\partial F^{(y)}}{\partial s_{kl}}}. \tag{2.4}$$

Der Tensor $\varkappa$ steht in Gleichung (2.4) für die Richtung der kinematischen Verfestigung. Er wird für das Übergangsfließpotential nach ZIEGLER [7] in Richtung von s festgelegt.

Die Grenzfläche (1) hingegen bezieht ihren Verfestigungszuwachs immer aus einer gemessenen monotonen Zug-(Druck-)kurve:

$$\mathrm{d}R^{(l)} = (1 - \xi^{(l)}) \left.\frac{\partial k_f}{\partial T}\right|_{\varepsilon_v^p = 0} \mathrm{d}T + \xi^{(l)} \left[\frac{\partial k_f}{\partial \varepsilon_v^p} \mathrm{d}\varepsilon_v^p + \frac{\partial k_f}{\partial T} \mathrm{d}T\right], \tag{2.5}$$

$$\mathrm{d}\zeta_{ij}^{(l)} = \mathrm{d}\zeta_{ij}^{(y)} - \mathrm{d}C\,\delta_{ij}. \tag{2.6}$$

Der Skalar $\mathrm{d}C$ in Gleichung (2.6) folgt aus der Forderung, daß sich Fließ- und Grenzfläche nicht überschneiden dürfen.

Folgen wir einmal beispielhaft einem schwellenden Ent- und Belastungszyklus im Modell: Bei Belastung aus dem Ausgangszustand folgt die Spannung zunächst der gemessenen nichtlinearen Fließkurve. Nach Entlastung und erneuter Belastung bis zur Plastifizierung erfährt die aktuelle Fließfläche wegen $E^p(\delta) \geqq \frac{\partial k_f}{\partial \varepsilon_v^p}$ eine größere kinematische Verfestigung als die Grenzfläche, so daß sie sich wieder der Grenzfläche bis zur Berührung nähert. Damit wird ein weicher Übergang von der elastischen Geraden zur monotonen Fließkurve möglich. Dies ist ein wichtiger Aspekt des Modells. Insgesamt kann mit diesem Ansatz die zyklische Ver- bzw. Entfestigung, die Hysterese und der Bauschinger-Effekt zweiter Art modelliert werden. Bild 3 zeigt dazu exemplarisch einen simulierten spannungs- und einen dehnungsgesteuerten Zyklus.

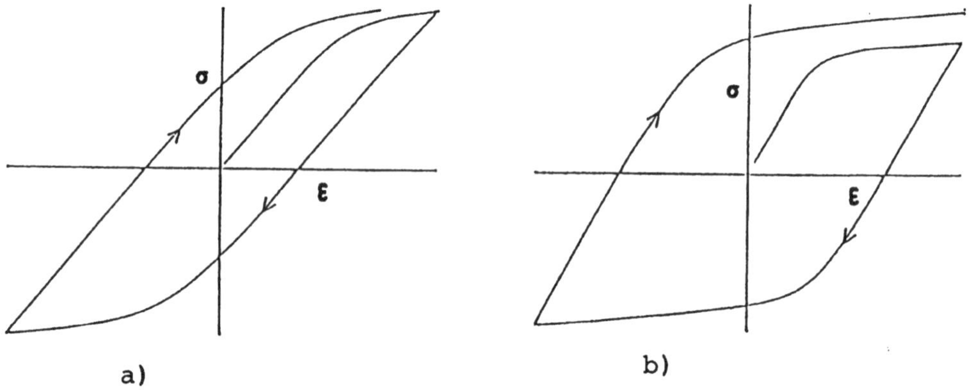

Bild 3. Simulierte Spannungs- (a) und dehnungsgesteuerte (b) Spannungs-Formänderungszyklen

Bauschinger-Effekte erster Art treten nach schwellender Belastung dann auf, wenn nach Entlastung des Werkstoffs wieder bis zur erneuten Plastifizierung aufbelastet wird. Der Plastifizierungsbeginn erfolgt dann bei einer geringeren Spannung, als sie am Ende der vorherigen Plastifizierung herrschte. Dieser Effekt kann durch die Zweiflächentheorie nicht ohne weiteres erfaßt werden. Daher wird innerhalb der Grenzfläche eine dritte, kleinere Fläche als untere Schranke angesetzt. Sie ist um genau den Betrag kleiner, der als Bauschinger-Effekt erster Art bei der aktuellen, plastischen Vergleichsdehnung maximal zulässig ist. Bei einer Folge von kleinen Entlastungsschritten werden diejenigen, die innerhalb des Bereichs zwischen Grenzfläche und dritter Potentialfläche verlaufen, mit einer isotropkinematischen Entfestigung behandelt. Weitere Entlastungsschritte unterhalb dieser dritten Fläche verlaufen dagegen – wie üblich – ohne Änderung der Fließfläche. Als Folge dieser anisotropen Entfestigung liegt nun der Fließbeginn bezüglich der alten Belastungsrichtung zwischen den beiden Potentialflächen – höchstens jedoch auf der dritten Fläche. Der Plastifizierungsbeginn in entgegengesetzter Richtung bleibt davon unberührt. Bei erneuter Belastung verschiebt sich nach einer verfrühten Plastifizierung die aktuelle Fließfläche mit isotrop-kinematischer Verfestigung wieder hin zur Grenzfläche.

An dieser Stelle sollte darauf hingewiesen werden, daß in der vorstehenden Betrachtung auch die „entfestigenden" Entlastungsschritte ohne zugehörige plastische Formänderung behandelt werden. Diese Annahme ist z. Z. üblich. Einige wenige Messungen mit großer Auflösung deuten jedoch darauf hin (vgl. auch Bild 1), daß die Entlastung bis zu der Spannung, ab der die Plastifizierung bei erneuter Aufbelastung eintritt, teilweise plastisch erfolgen kann. Ebenfalls erkennt man, daß auch während der Entlastung die Ansatzwerte α von Null auf positive Werte ansteigen können. Dies läßt vermuten, daß die aktuelle Fließfläche bei Entlastung von einem Zylinder in ein Rotationsellipsoid übergeht. Der Wechsel der Fließflächengeometrie erfolgt somit in umgekehrter Folge wie beim Belastungsschritt. Sollte sich diese Vermutung experimentell absichern lassen, so müßte das Modell für Be- und Entlastung modifiziert werden.

Rißmodell

Eingangs wurde schon darauf hingewiesen, daß die spröden (elastischen) Reaktionsschichten an den Phasengrenzen zwischen den Fasern und der Matrix außerordentlich sensibel sind. Kleine Anrisse, die vom Fertigungsprozeß her oft unvermeidbar sind, können sich dort rasch ausbreiten und so zu einer teilweisen Ablösung der Fasern führen. Dies könnte wiederum Rückwirkungen auf die werkstoffmechanische Stabilität der Matrix bei zyklischer Belastung haben.

Die numerische Simulation solcher Risse erfolgt über die Finite Elemente Methode. Hierzu wird aus rechentechnischen Gründen für die thermomechanische und für die bruchmechanische Simulation ein gemeinsames FE-Netz mit relativ hohem Diskretisierungsgrad verwendet. Für den bruchmechanischen Teil wird dabei eine Vorgehensweise weiterentwickelt, die als sogenanntes „smeared crack-“ oder „fictious crack-model“ bekannt geworden ist [8, 9, 10]. Die Idee des Ansatzes besteht darin, daß beim Erreichen des Bruchkriteriums (unter Beibehaltung der Struktur des FE-Netzes) unter Zugspannungen senkrecht zum Riß die Steifigkeit des Rißelements für diese Richtung auf einen Wert nahe Null gesetzt wird. Damit können Zug- und Druckspannungen in der Rißebene weiterhin übertragen werden. Senkrecht dazu sind dagegen nur Druckspannungen und (vernachlässigbar) kleine Zugspannungen möglich. Risse mit großen Unterschieden in den Verschiebungen benachbarter Punkte der Rißoberflächen können so nicht modelliert werden. Sie treten in der späteren Fallstudie aber auch nicht auf.

Die orthotrope Matrix der linear-elastischen Konstanten lautet für ein Rißelement bei ebener Formänderung:

$$\boldsymbol{D} = \frac{1}{1 - \nu_1 - 2\frac{E_1}{E_2}\nu_2^2} \begin{bmatrix} \frac{E_1}{1+\nu_1}\left(1 - \frac{E_1}{E_2}\nu_2^2\right) & E_1\nu_2 & 0 \\ E_1\nu_2 & E_2(1-\nu_1) & 0 \\ 0 & 0 & G_2\left(1 - \nu_1 - 2\frac{E_1}{E_2}\nu_2^2\right) \end{bmatrix} \tag{2.7}$$

mit E — Elastizitätsmodul, ν — Querzahl und G — Schubmodul.

Für die isotrope Rißbildungsebene (1,3-Ebene) gilt $E_1 = E_3$ und $\nu_1 = \nu_3$. Gemäß Voraussetzung liegt ebene Formänderung in der 1,2-Ebene vor.

Da in der Bruchfläche keine Schubspannungen herrschen sollen, fallen Riß- und Hauptspannungsrichtung zusammen. Die Matrix (2.7) ist daher in das globale Koordinatensystem zu transformieren. Tritt nach einem Lastschritt Zugspannung senkrecht zum Riß auf, dann wird im neu gerissenen Element die Steifigkeit für diese Richtung bis nahe Null herabgesetzt. Damit läßt sich der Riß gut simulieren.

Als Bruchkriterium für die hier vorliegende Mixed-Mode-Beanspruchung dient das Kriterium der Energiefreisetzungsrate [11]. Sie ist bei Mode I- und Mode II-Belastung für Rißausbreitung in Richtung φ gegeben durch

$$G = \cos\varphi \frac{1-\nu^2}{E}(K_I^2 + K_{II}^2) - \sin\varphi \frac{2(1-\nu^2)}{E} K_I K_{II}. \tag{2.8}$$

K_I, K_{II} — Spannungsintensitätsfaktoren für Mode I und II.

Mit dem Extremwert von G in Gleichung (2.8) ist auch die Rißausbreitungsrichtung φ_0 festgelegt.

Vergleicht man die maximale Energiefreisetzungsrate G_{max} mit dem kritischen Wert

$$G_{Ic} = \frac{1-\nu^2}{E} K_{Ic}^2, \tag{2.9}$$

(K_{Ic} — Rißzähigkeit für Mode I) so liegt ein Bruchkriterium vor, das auch später in der numerischen Rechnung verwendet wird.

Temperaturverteilung

Die problembeschreibende nichtlineare, instationäre Wärmeleitungsgleichung lautet:

$$\nabla_i(k_{ij}(T)\,\nabla_j T) + Q(t) - c(T)\,\varrho(T)\,\dot{T} = 0, \tag{2.10}$$

mit $k_{ij}(T)$ — Wärmeleitfähigkeitsmatrix, $Q(t)$ — innere Wärmeleistung, $c(T)$ — Wärmekapazität und $\varrho(T)$ — Dichte.

Auf dem beliebig stückweise definierten Rand der Struktur können folgende Randbedingungen vorgegeben werden:

— Wärmeübergang an ein angrenzendes, konvektierendes Medium mit zeitlich variierender Temperatur $T_\infty(t)$:

$$q_i n_i = h(T)\,[T - T_\infty(t)],$$

mit $q_i n_i$ — Wärmestrom senkrecht zur Oberfläche und $h(T)$ — Wärmeübergangszahl.

— Wärmestrom durch Differenz von Eigen- und Fremdstrahlung eines benachbarten Körpers mit zeitlich variierender Temperatur $T_\infty(t)$:

$$q_i n_i = \beta\,\epsilon(T)\,[T^4 - T_\infty^4(t)],$$

mit β — Stefan-Boltzmann-Konstante und $\epsilon(T)$ — Emissivität.

— Randtemperatur auf einem beliebig großen Teil der Randfläche.

Damit sind die wichtigsten Grundlagen zur Bearbeitung der späteren Fallstudie angegeben. Hinreichend bekannte Zusammenhänge wurden dabei nur kurz angesprochen.

3. Lösungsverfahren

Da die FEM als Standard-Verfahren der numerischen Mathematik hinreichend bekannt ist, sollen hier nur problemorientierte Punkte angesprochen werden. Grundsätzlich werden aus Gründen der Speicherverwaltung und der Rechengeschwindigkeit die thermischen und die mechanischen Rechnungen mit einem gemeinsamen Diskretisierungsnetz durchgeführt. Zwar stimmen die Bereiche mit großen Spannungsgradienten und dem dort wünschenswerten engen Netz nicht notwendigerweise mit Bereichen großer Temperaturgradienten überein, doch kann dieser Mangel durch einen insgesamt hohen Diskretisierungsgrad behoben werden.

Mit der Integralformulierung der Wärmeleitungsgleichung (2.10) und der Einführung der Strahlungs- und Konvektionsrandbedingungen kann ein Differentialgleichungssystem entwickelt werden:

$$C_{rs}(T) \cdot \dot{T}_s + L_{rs}(T) \cdot T_s = F_r^Q(t) + F_r^h(t, T) + F_r^R(t, T), \tag{3.1}$$

mit C_{rs} – Kapazitätsmatrix, L_{rs} – Leitfähigkeitsmatrix, T_s – Knotentemperaturvektor, F_r^Q – Wärmequellenvektor, F_r^h – Konvektionslastvektor, F_r^R – Strahlungslastvektor.

Die zeitliche Integration wird mit einem modifizierten Crank-Nicholson Schema durchgeführt. Dieses halbimplizite Verfahren bietet den Vorteil großer Stabilität des Algorithmus. Zu große Zeitschritte können allerdings zu starken Schwingungen im zeitlichen Verlauf der Lösung führen. Deshalb kommt eine Abschätzung für die kritische Zeitschrittgröße nach MYERS [12] zur Anwendung. Bei Schrittweiten unterhalb dieses abgeschätzten kritischen Wertes kann die Lösung als sicher angenommen werden.

Der Wärmequellenlastvektor in Gleichung (3.1) berücksichtigt die Wärmeleistung der plastischen Formänderung, jedoch jeweils aus dem vorhergehenden Lastschritt. Da die Schrittgröße wegen der Nichtlinearität des Problems sehr klein gehalten werden muß, ist der Fehler durch die verspätete Zuführung der Wärme aus der plastischen Formänderung vertretbar gering.

Die Finite Elemente Formulierung für den inelastischen Teil des Problems folgt aus dem Prinzip der virtuellen Arbeit. Die Rechnung wird dabei als Erweiterung der elastischen Analyse aufgefaßt. Für die virtuelle Arbeit gilt im Fall des mechanischen Gleichgewichts

$$\delta W = \int_V \sigma_{ij} \delta\varepsilon_{ij} \, dV - \int_A f_i^A \delta v_i^A \, dA = 0. \tag{3.2}$$

Mit f_i^A – Koordinaten der Spannungen an der Oberfläche, δv_i^A – Koordinaten der virtuellen Randverschiebungen, V – Volumen und A – Oberfläche.

Schreibt man für die Spannungskoordinaten

$$\sigma_{ij} = D_{ijkl}(\varepsilon_{kl} - \varepsilon_{kl}^p - \varepsilon_{kl}^{th}) \tag{3.3}$$

und entwickelt damit Gleichung (3.2), so folgt daraus nach Summation über alle Elemente das Gesamtgleichungssystem:

$$K_{rs}(T)\, u_s = F_r^p(\varepsilon_v^p, T) + F_r^{th}(T) + F_r^A(t). \tag{3.4}$$

In den Gleichungen (3.2) und (3.3) bedeuten D_{ijkl} – elastische Konstanten, $K_{rs}(T)$ – Steifigkeitsmatrix, u_s – Knotenverschiebungen, F_r^p – plastischer Lastvektor, F_r^{th} – thermischer Anfangslastvektor, F_r^A – äußerer Lastvektor.

Die bruchmechanische Rechnung wird über einen Bruchlastvektor eingeführt. Lautet zum Beispiel die Gleichung für einen beliebigen Freiheitsgrad n

$$K_{ns} u_s = F_n$$

herausgegriffen aus dem Gleichungssystem (3.4) und ändert sich durch das Versagen eines Elements der Steifigkeitskoeffizient K_{nm} nach K_{nm}^* und die Lastvektoren von F_n nach F_n^*, so wird, statt die globale Steifigkeitsmatrix neu aufzubauen, der folgende Bruchlastvektor zu den Kräften F_n hinzuaddiert:

$$F_n^b = (K_{nm} - K_{nm}^*)\, u_m + (F_n^* - F_n). \tag{3.5}$$

Damit erhält man das Gesamtgleichungssystem der gerissenen Struktur ohne Neuformulierung der Gesamtsteifigkeitsmatrix.

Parallel zu den elastisch-plastischen Rechnungen wird zu jedem Zeitpunkt die aktuelle Temperaturverteilung der Struktur nach Gleichung (3.1) berechnet. Somit können die durch Temperaturänderung hervorgerufenen thermischen Dehnungen über den thermischen Anfangslastvektor in den Gesamtlastvektor mit aufgenommen werden. Nachdem das Gleichungssystem (3.4) gelöst ist, wird mit den neuen Verschiebungsinkrementen der Spannungszustand für jedes Element berechnet und das Fließkriterium überprüft. Für alle plastischen Elemente folgt dann nach einer Subinkrementierung des aktuellen Spannungszuwachses die elastisch-plastische Rechnung. Da das hier verwendete vollexplizite Integrationsverfahren die nichtlineare Spannung nicht exakt trifft, muß der Konsistenzfehler $dF \neq 0$ in einer Iteration bis unter eine Genauigkeitsschranke verringert werden. Die so ermittelte Differenz $\Delta\sigma^p$ zwischen dem elastisch vorausberechneten Spannungszuwachs $\Delta\sigma^e$ und dem iterierten Zuwachs $\Delta\sigma$ geht in den plastischen Pseudolastvektor ein. Mit diesem Vektor wird nun für einen festen Zeitpunkt das Gesamtgleichungssystem in einem modifizierten Newton-Raphson-Verfahren gelöst.

Weiterhin werden die nach der elastisch-plastischen Rechnung ermittelten dissipierten Umformleistungen in den Wärmequellenlastvektor des Temperaturgleichungssystems zurückgeführt. Damit ist die thermomechanische Kopplung erfolgt.

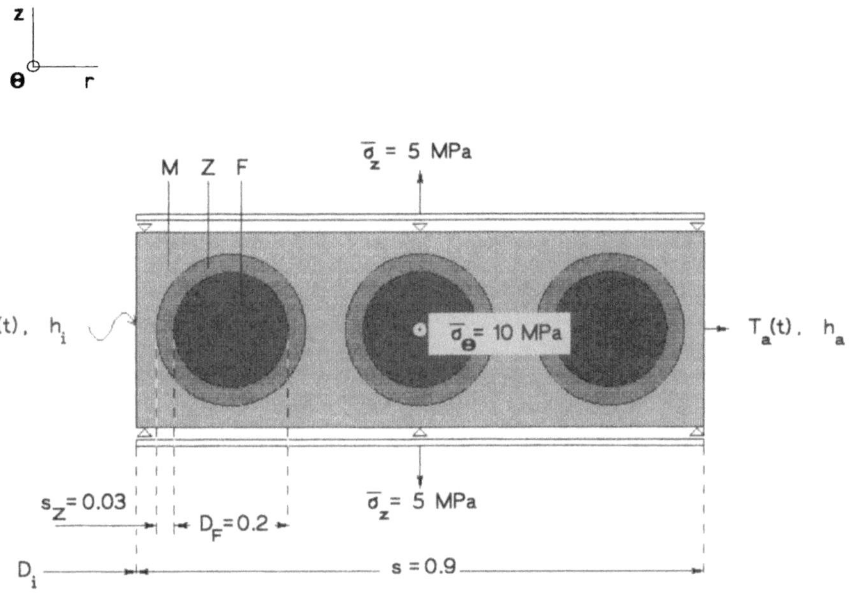

Bild 4. Faserverbundrohr-Wandquerschnitt mit thermomechanischer Belastung

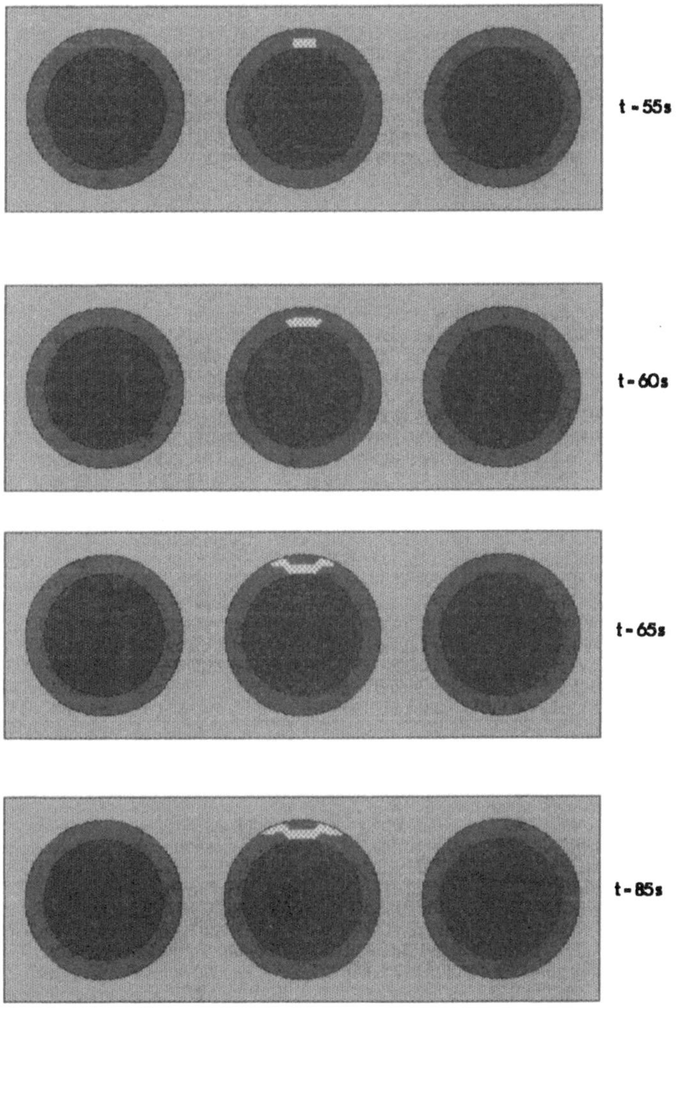

Bild 5. Rißausbreitung bei horizontalem Anriß

4. Fallstudie

Exemplarisch für viele technische Strukturen wird die werkstoffmechanische Stabilität eines metallischen Faserverbundrohres bei zyklischer thermomechanischer Belastung numerisch untersucht. Der Rohrwandquerschnitt ist in Bild 4 dargestellt. Das Rohr hat einen Innendurchmesser $D_i \gg 2s$ und eine Wandstärke $s = 0{,}9$ mm. Die Fasern (F) (X2NiCoMo, $D_F = 0{,}2$ mm) sind in der Matrix (M) (AlMgSi1) quadratisch angeordnet, d. h. alle Fasern haben gleiche Abstände zueinander. Sie sind von einer elastischen (spröden) Reaktionsschicht (Z) (Al_5Fe_2, $s_z = 0{,}03$ mm) umgeben. Das Rohr wird in Achsrichtung durch die mittlere Spannung $\bar{\sigma}_z = 5$ MPa belastet. In Umfangsrichtung (Faserrichtung) wird eine mittlere Spannung $\bar{\sigma}_\Theta = 10$ MPa vorgegeben. Im Innern des Rohres befinde sich eine Flüssigkeit mit periodisch sich ändernder Temperatur

$$T_i(t) = 40 \sin\left(\frac{\pi}{100} t - \frac{\pi}{2}\right) + 40 \,[^\circ\mathrm{C}],\ t[\mathrm{s}]\,.$$

An der Rohraußenfläche herrsche die konstante Temperatur $T_a = 20$ °C. Die Wärmeübergangszahlen betragen innen $h_i = 0{,}5$ W/cm^2 K und außen $h_a = 0{,}6 \cdot 10^{-6}$ W/cm^2 K.

Die Elastizitätsmoduli der Verbundphasen betragen für 0 °C: $E_M = 70$ GPa, $E_z = 134$ GPa, $E_F = 213$ GPa. Die Anfangsfließspannung (Anfangsformänderungsfestigkeit) ist für die Matrix bei 0 °C: $k_{fO} = 80$ MPa. Auf die Darstellung aller werkstoffmechanischen Kenngrößen in Abhängigkeit von der Temperatur wird aus Gründen der Platzersparnis verzichtet.

Die numerische Simulation der thermomechanischen Belastung der ungeschädigten Rohrwand in Bild 4 zeigt, daß die Struktur nach etwa 10 Zyklen (2000 s) stabilisiert ist. Danach treten bei weiteren Lastzyklen keine inelastischen Formänderungen in der Matrix mehr auf.

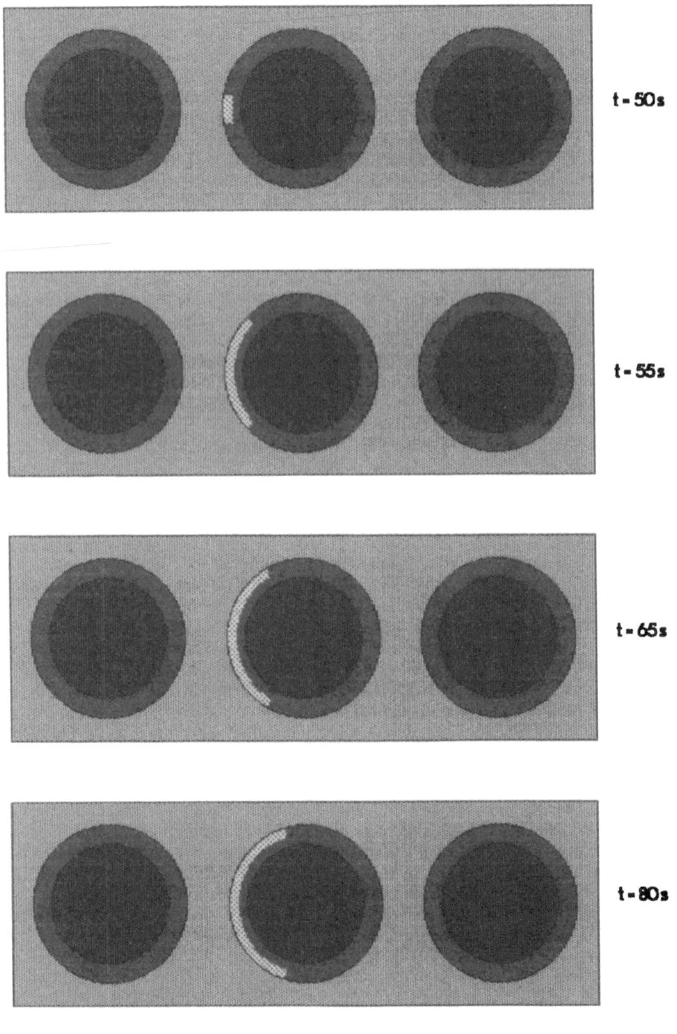

Bild 6. Rißausbreitung bei vertikalem Anriß

Von größerem Interesse ist das Verhalten einer geschädigten Struktur. Dazu wird an der horizontalen Phasengrenze (F/Z) der mittleren Faser ein 0,03 mm langer Riß vorgegeben und die Struktur wieder wie oben angegeben belastet. In der Bildfolge 5 erkennt man, daß sich der Riß zunächst an der Phasengrenze (F/Z) ausdehnt und sich dann zur Phasengrenze (Z/M) hin ausbreitet. Nach $t = 80$ s kommt der Riß zum Stillstand. Die Gesamtstruktur ist wiederum nach etwa 10 Zyklen stabilisiert.

In der Bildfolge 6 wird nun an der vertikalen Phasengrenzfläche (Z/M) ein 0,03 mm langer Riß vorgegeben. Man sieht, daß sich der Riß entlang der Phasengrenze (Z/M) ausbreitet und nach $t = 80$ s ebenfalls zum Stillstand kommt. Auch hier hat sich die Struktur nach etwa 10 Zyklen stabilisiert.

Bild 7a zeigt das Spannungsfeld $\sigma_z(t)$ zur Zeit $t = 100$ s bei horizontalem Riß (vgl. Bild 5). Während das Spannungsfeld in der Umgebung der linken und rechten Faser weitgehend dem der ungeschädigten Struktur entspricht, geht vom Riß an der Phasengrenzfläche (F/Z) der mittleren Faser in der Matrix eine Spannungsabsenkung aus. Der Riß endet dort, wo in der ungeschädigten Struktur die Spannung σ_z klein ist. Gleiches gilt für den vertikalen Riß in Bild 6. Hier ist die treibende Kraft für den Riß die Spannung σ_r. Der Riß endet wiederum an der Stelle, wo die Spannung σ_r der ungeschädigten Struktur klein wird. Bild 7b zeigt das Spannungsfeld $\sigma_r(t)$ bei vertikalem Riß nach 100 s.

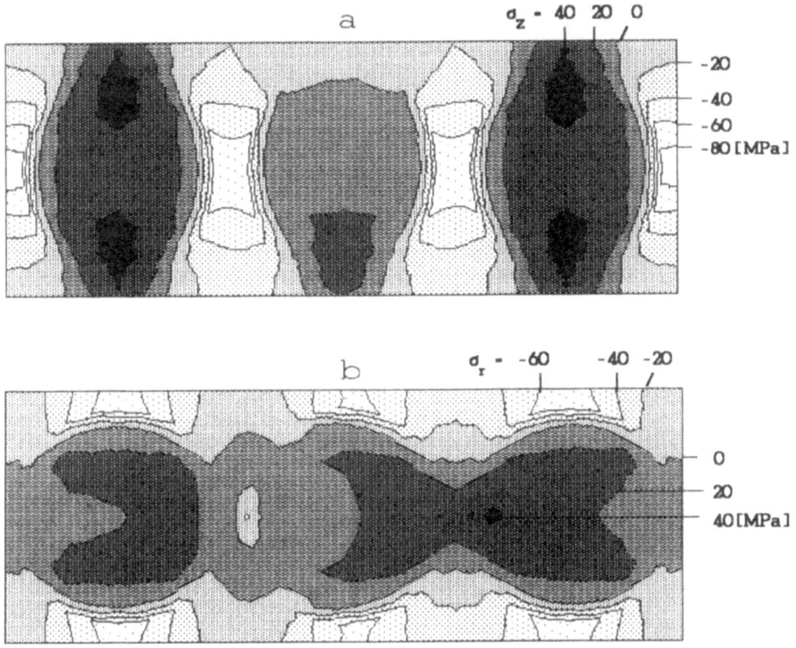

Bild 7. Spannungsverteilung bei horizontalem (a) und vertikalem (b) Anriß zur Zeit $t = 100$ s

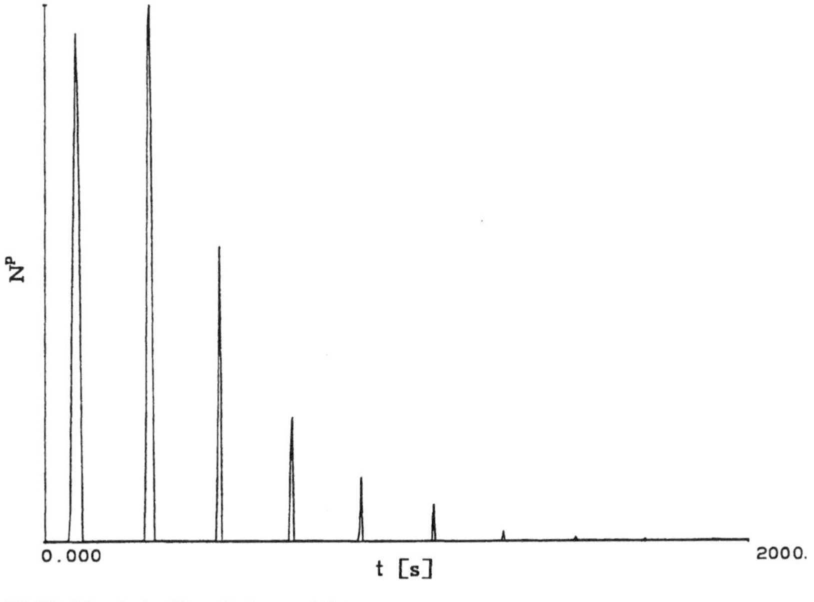

Bild 8. Plastische Formänderungsleistung

In Bild 8 wird schließlich die Entwicklung der plastischen Formänderungsleistung wiedergegeben. Man sieht, daß sich die Struktur nach etwa 10 Zyklen werkstoffmechanisch stabilisiert hat. Zwischen der ungeschädigten und den geschädigten Strukturen zeigen sich keine nennenswerten Unterschiede in der Entwicklung der plastischen Formänderungsleistung der Matrix. Die Fließgrenze der Fasern wird nicht erreicht.

Für die vorliegenden Fallstudien läßt sich feststellen, daß herstellungsbedingte kleine Anrisse in einer spröden Übergangsschicht zwischen Faser und Matrix, so wie sie hier angenommen wurden, weniger kritisch für die werkstoffmechanische Stabilität sind, als man vermuten könnte. Diese Risse hatten weder Einfluß auf das Einspielverhalten der plastifizierten Matrix noch verursachten sie ein katastrophales Auseinanderbrechen der gesamten Struktur.

Die Autoren danken der Deutschen Forschungsgemeinschaft für ihre Unterstützung

Literatur

1 Hashin, Z.: The elastic moduli of heterogeneous materials. J. Appl. Mech. **29** (1962), 143–150.
2 Hashin, Z.: Analysis of composite materials. J. Appl. Mech. **50** (1983), 481–505.
3 Reuss, A.: Berücksichtigung der elastischen Formänderung in der Plastizitätslehre. ZAMM **10** (1930), 266–274.
4 Mahrenholtz, O.; Ismar, H.: Zum elastisch-plastischen Übergangsverhalten metallischer Werkstoffe. Ing. Arch. **50** (1981), 217–224.
5 Ismar, H.; Schmitt, J.: Zur zyklischen Belastung einer Aluminiumlegierung im Bereich kleiner plastischer Formänderungen. ZAMM **70** (1990) 4, T 324–T 327.
6 Dafalias, Y. F.; Popov, E. P.: Plastic internal variables formalism of cyclic plasticity. J. Appl. Mech. **43** (1976), 645–654.
7 Ziegler, H.: A modification of Pragers hardening rule. Quart. Appl. Math. **17** (1959), 55–65.
8 de Borst, R.: Smeared cracking, plasticity, creep and thermal loading – a unified approach. Comp. Meth. Appl. Mech. Eng. **62** (1987), 89–110.
9 Hillerborg, A.: Analysis of one single crack. In: Wittmann: Fracture mechanics of concrete. Elsevier, Amsterdam 1983, pp. 223–249.
10 Sadouki, H.; Wittmann, F. H.: On the analysis of the failure process in composite materials by numerical simulation. Materials Science and Engineering **104** (1988), 9–20.
11 Strifors, H. C.: A generalized force measure of conditions at crack tips. Int. J. Solids Structures **10** (1974), 1389–1404.
12 Myers, G. E.: The critical time step for FE solutions to two – dimensioned heat-conduction transients. Transactions of the ASME **100** (1978), 120–127.

Received September 7, 1990

Anschriften: Dr.-Ing. Rudolf Dressler, Prof. Dr.-Ing. Heinz Ismar, Dipl.-Ing. Wolfgang Ripplinger, Lehrstuhl für Technische Mechanik, Universität des Saarlandes, W-6600 Saarbrücken, Deutschland

ZAMM · Z. angew. Math. Mech. 71 (1991) 7/8, 277–288 Akademie Verlag

KAMPOWSKY, W.

Optimal Control in Nonlinear Evolution Equations with Constraints

Dedicated to Professor Dr. L. BITTNER at the occation of his 60th birthday

Die Arbeit setzt die Untersuchungen des Autors [5]–[12] zur optimalen Steuerung von nichtlinearen Evolutionsgleichungen mit monotonen Operatoren fort. Es werden notwendige Optimalitätsbedingungen hergeleitet, wobei zusätzliche Nebenbedingungen allgemeiner Gestalt für die Trajektorie und die Steuerung zugelassen sind. In der Anwendung wird ein Dirichlet-Randsteuerproblem und ein zeitoptimales Problem betrachtet.

This paper continues the investigations of the author [5]–[12] concerning optimal control of nonlinear evolution equations with monotone operators. Necessary optimality conditions are derived where additional constraints in a general form for the state and the control are allowed. As applications a Dirichlet boundary control problem and a time-optimal problem are considered.

Настоящая работа продолжает исследования автора [5]–[12] посвящено оптимальному управлением нелинейных эволюционных уравнений с монотонными операторами. Необходимые условия оптимальности установлены, причем дополнительные условия общего типа для траектория и для управления оставлены. В применении задача граничего управлении Дирихле и задача оптимального быстродействия рассматриваются.

MSC (1980): 90C30

Introduction

In earlier papers [5]–[12] we have considered in detail optimal control problems governed by nonlinear initial-boundary value problems for parabolic differential equations of second order of divergence type. The typical equation of this class is the wellknown heat equation or more general, the diffusion equation.

The most essential aspect of classification is the following one. We solve the initial-boundary value problem for a parabolic differential equation as an evolution equation using the theory of monotone operators (see LIONS [14], GAJEWSKI, GRÖGER, ZACHARIAS [4]). Applying this way of definition of generalized solutions for initial-boundary value problems we are able to deal with nonlinear problems, especially, with a nonlinear main part of the parabolic differential equation. In this paper we consider additional constraints in a general form for the state as well, as for the control. In ch. 1 we formulate an abstract model of necessary optimality conditions which is due to BITTNER [1]–[3] including two modifications. We concrete this model for control problems governed by evolution equations with constraints in ch. 2. Finally, we close in ch. 3 with two examples, at first, a problem of boundary control in a Dirichlet condition, and secondly, a time-optimal control problem.

1. On BITTNER's model of necessary optimality conditions

Let us consider the following problem of optimal control:

$$\left.\begin{array}{l} F(x, u) = \min! \\ G(x, u) = 0\,, \qquad H(x, u) \in Z_0\,, \qquad u \in U_0\,, \end{array}\right\} \tag{1}$$

where $F\colon W \times U \to \mathbb{R}$, $G\colon W \times U \to Y$, $H\colon W \times U \to Z$, W, Y and Z are Banach spaces, $Z_0 \subseteq Z$, $U_0 \subseteq U$ and U is an arbitrary set.

There are two differences between the original model of BITTNER [1]–[3] and this modified version.

We will generally suppose the existence of the derivatives with respect to $x \in W$ only in the sense of GÂTEAUX, not in the sense of FRÉCHET. (We remember that a mapping from L_2 into L_2 pointwise given by a nonlinear Carathéodory function has no Fréchet derivative.) Furthermore, we consider additional constraints $H(x, u) \in Z_0$ where $Z_0 \subseteq Z$ and Z may be an infinite-dimensional space.

Let (x_0, u_0) be optimal for problem (1). We suppose that F, G and H have Gâteaux derivatives $F_x(x, u)$, $G_x(x, u)$ and $H_x(x, u)$ with respect to $x \in W$ for all x of a neighbourhood of x_0 and for all $u \in U_0$.

We consider variations of the optimal control u_0 in the following sense. Let $\{u_k\} \subseteq U_0$ and let us suppose that $F_x(., u_k)\, x\colon W \to \mathbb{R}$ and $H_x(., u_k)\, x\colon W \to Z$ are continuous for all $x \in W$ uniformly with respect to u_k and that

$$F_x(x_0, u_k)\, x \to F_x(x_0, u_0)\, x \quad \text{and} \quad H_x(x_0, u_k)\, x \to H_x(x_0, u_0)\, x \quad \text{for all} \quad x \in W\,.$$

We call such a sequence $\{u_k\}$ an *admissible sequence of varied controls for* (x_0, u_0).

Further, we define the following directional limits: Let $\{\gamma_k\} \subseteq \mathbb{R}_+$ with $\gamma_k \downarrow 0$ and let $\{u_k\} \subseteq U_0$. Then

$$\Delta F = \lim_{k\to\infty} \frac{1}{\gamma_k} (F(x_0, u_k) - F(x_0, u_0)), \qquad \Delta G = \lim_{k\to\infty} \frac{1}{\gamma_k} (G(x_0, u_k) - G(x_0, u_0)),$$

$$\Delta H = \lim_{k\to\infty} \frac{1}{\gamma_k} (H(x_0, u_k) - H(x_0, u_k))$$

denote the directional limits of F, G and H with respect to u at the point (x_0, u_0) which are generated by $\{\gamma_k\}$ and $\{u_k\}$.

Now we formulate general assumptions concerning the solution of the process equation $G(x, u) = 0$.

Let $\{\gamma_k\} \subseteq \mathbb{R}_+$ with $\gamma_k \downarrow 0$, let $\{u_k\} \subseteq U_0$ and let ΔG exist. We assume that:

For all $k \gg 1$ there exists $x_k \in W$ such that $G(x_k, u_k) = 0$. (2.1)

The directional limit $\Delta x = \lim_{k\to\infty} \frac{1}{\gamma_k} (x_k - x_0)$ exists. (2.2)

Δx is a solution of the variational process equation $G_x(x_0, u_0)\,\Delta x = -\Delta G$. (2.3)

Remark 1: Suppose that $G_x(., u_k)\,x \colon W \to Y$ is continuous for all $x \in W$ uniformly with respect to u_k and that $G_x(x_0, u_k)\,x \to G_x(x_0, u_0)\,x$ for all $x \in W$. Then (2.1) and (2.2) imply (2.3).

Remark 2: A general realization of the assumptions (2.1)–(2.3) can be constructed using the wellknown fixpoint theorem of Banach. Indeed, the following assertion is valid (see BITTNER [1]–[3]).

Suppose that $G_x(x, u)$ even exists as a Fréchet derivative for all x of a neighbourhood of x_0 and for $u \in U_0$ and that $G_x(x_0, u_0)$ is a one-to-one mapping. Further, suppose that $G_x(., u_k)\colon W \to \mathscr{L}(W; Y)$ is continuous uniformly with respect to u_k and that $G_x(x_0, u_k) \to G_x(x_0, u_0)$. Then the assumptions (2.1)–(2.3) are fulfilled.

We consider the set Φ of all common directional limits $(\Delta F, \Delta G, \Delta H)$ which are generated by such sequences $\{\gamma_k\} \subseteq \mathbb{R}_+$ with $\gamma_k \downarrow 0$ and such admissible sequences $\{u_k\}$ of varied controls for (x_0, u_0) that the assumptions (2.1)–(2.3) are fulfilled.

The set Φ is a cone in $\mathbb{R} \times Y \times Z$. To derive necessary optimality conditions we need a convex subcone Φ_0 of Φ. In all investigated cases it was popssible to find such convex subcones.

We define the following subsets in $\mathbb{R} \times Z$:

$$\left.\begin{aligned} K &= \{(F_x(x_0, u_0)\,\Delta x + \Delta F, H_x(x_0, u_0)\,\Delta x + \Delta H) \mid (\Delta F, \Delta G, \Delta H)) \in \Phi_0, G_x(x_0, u_0)\,\Delta x = -\Delta G\},\\ L &= \{(\xi, z) \mid \xi < 0, z \in \operatorname{int} Z_0 - H(x_0, u_0)\}. \end{aligned}\right\} \quad (3)$$

If $\Phi_0 \subseteq \Phi$ is a convex cone then also K, and if $Z_0 \subseteq Z$ is a convex set with a nonempty interior then also L.

Theorem 1: *Let $Z_0 \subseteq Z$ be a convex set with a nonempty interior. Then the relation $K \cap L = \emptyset$ follows from the optimality of (x_0, u_0).*

Proof: Suppose the contrary: $(\xi, z) \in K \cap L$. The following relations hold:

$$\xi = F_x(x_0, u_0)\,\Delta x + \Delta F < 0, \quad (4.1)$$

$$z = H_x(x_0, u_0)\,\Delta x + \Delta H = z_0 - H(x_0, u_0). \quad (4.2)$$

Since $z_0 \in \operatorname{int} Z_0$ there exists a neighbourhood N of the origin in Z such that $z_0 + N \subseteq Z_0$. Since $H(x_0, u_0) \in Z$ we get from the convexity of Z_0 that

$$\gamma z_0 + \gamma N + (1 - \gamma)\,H(x_0, u_0) \subseteq Z_0 \quad \text{for} \quad \gamma \in [0, 1]. \quad (4.3)$$

The definitions of Δx, ΔF and ΔH imply

$$\left.\begin{aligned} x_k - x_0 &= \gamma_k\,\Delta x + o_x(\gamma_k),\\ F(x_0, u_k) - F(x_0, u_0) &= \gamma_k\,\Delta F + o_F(\gamma_k),\\ H(x_0, u_k) - H(x_0, u_0) &= \gamma_k\,\Delta H + o_H(\gamma_k). \end{aligned}\right\} \quad (4.4)$$

For $k \gg 1$ the process equation $G(x_k, u_k) = 0$ is valid.

First, we prove that $H(x_k, u_k) \in Z_0$ for $k \gg 1$. – Using the mean value theorem and (4.2)–(4.3) we calculate

$$\begin{aligned} H(x_k, u_k) &= H(x_k, u_k) - H(x_0, u_k) - H_x(x_0, u_k)(x_k - x_0) \\ &\quad + [H_x(x_0, u_k) - H_x(x_0, u_0)](x_k - x_0) + H_x(x_0, u_0)(x_k - x_0) \\ &\quad + [H(x_0, u_k) - H(x_0, u_0)] + H(x_0, u_0) \\ &= \gamma_k \left\{ \int_0^1 [H_x(x_0 + \tau(x_k - x_0), u_k) - H_x(x_0, u_k)] \frac{1}{\gamma_k}(x_k - x_0)\, d\tau \right. \\ &\quad \left. + [H_x(x_0, u_k) - H_x(x_0, u_0)] \frac{1}{\gamma_k}(x_k - x_0) + H_x(x_0, u_0) \frac{1}{\gamma_k} o_x(\gamma_k) + \frac{1}{\gamma_k} o_H(\gamma_k) \right\} \\ &\quad + \gamma_k (H_x(x_0, u_0)\, \Delta x + \Delta H) + H(x_0, u_0) \\ &= \gamma_k \{\dots\} + \gamma_k z_0 + (1 - \gamma_k) H(x_0, u_0) \in Z_0 \quad \text{for} \quad k \gg 1 \end{aligned}$$

since $\{\dots\} \in N$.

Furthermore, as above we obtain $F(x_k, u_k) = \gamma_k\{\dots\} + \gamma_k(F_x(x_0, u_0)\,\Delta x + \Delta F) + F(x_0, u_0)$ where $\{\dots\} \in N$. The contradiction $F(x_k, u_k) < F(x_0, u_0)$ for $k \gg 1$ follows from (4.1) which completes the proof. ■

Now we apply a wellknown separation theorem to get necessary optimality conditions.

Theorem 2: *Let $\Phi_0 \subseteq \Phi$ be a convex cone and let $Z_0 \subseteq Z$ be a convex set with a nonempty interior. Then a necessary optimality condition for (x_0, u_0) to be optimal is the following one:*

There exist multipliers $\lambda \geqq 0$ and $z^* \in Z^*$ (λ and z^* not both zero) such that there hold the variational inequality

$$\langle z^*; H_x(x_0, u_0)\,\Delta x + \Delta H \rangle + \lambda(F_x(x_0, u_0)\,\Delta x + \Delta F) \geqq 0 \quad \textit{for all} \quad (\Delta F, \Delta G, \Delta H) \in \Phi_0\,, \tag{5.1}$$

where Δx is a solution of the variation equation

$$G_x(x_0, u_0)\,\Delta x = -\Delta G\,, \tag{5.2}$$

and

the maximum condition

$$\langle z^*; H(x_0, u_0) \rangle = \max_{z \in Z_0} \langle z^*; z \rangle\,. \tag{5.3}$$

Proof: Applying a separation theorem there exist $\lambda \in \mathbb{R}$, $z \in Z$ (not both zero) and $\alpha \in \mathbb{R}$ such that

$$\lambda\xi + \langle z^*; z \rangle \geqq \alpha \quad \text{for all} \quad (\xi, z) \in K\,, \tag{6.1}$$

$$\lambda\xi + \langle z^*; z \rangle \leqq \alpha \quad \text{for all} \quad (\xi, z) \in L\,. \tag{6.2}$$

Obviously, $(0, 0) \in K$ which implies $\alpha \leqq 0$.

$H(x_0, u_0) \in Z_0$, but we can find a sequence $\{z_k\} \subseteq \operatorname{int} Z_0$ which tends to $H(x_0, u_0)$. If $\{\xi_k\} \subseteq \mathbb{R}_-$ tends to zero then $(\xi_k, z_k - H(x_0, u_0)) \in L$. Using (6.2) we get $\alpha = 0$.

Evidently, $\lambda \geqq 0$ follows from (6.2), too. – If $\xi < 0$ and $z \in \operatorname{int} Z_0$ are arbitrary then relation (6.2) implies $\langle z^*, z \rangle \leqq \langle z^*; H(x_0, u_0) \rangle$ and therefore, the maximum relation (5.3). The variational inequality (5.1) follows directly from (6.1) by the definition of K. ■

We define an adjoint equation belonging to the optimal solution (x_0, u_0),

$$G_x^*(x_0, u_0)\, y^* = w^*\,, \qquad y^* \in Y^*\,, \tag{7}$$

where $w^* \in W^*$ is given.

Corollary: *If $y^* \in Y^*$ is a solution of the following adjoint equation,*

$$G_x^*(x_0, u_0)\, y^* = \lambda F_x(x_0, u_0) + H_x^*(x_0, u_0)\, z^*\,, \tag{8.1}$$

then the variational inequality (5.1) can be written as follows:

$$\langle z^*; \Delta H \rangle - \langle y^*; \Delta G \rangle + \lambda\, \Delta F \geqq 0 \quad \textit{for all} \quad (\Delta F, \Delta G, \Delta H) \in \Phi_0\,. \tag{8.2}$$

2. Optimal control in evolution equations

2.1. The problem, assumptions and preparing assertions

Let V be a reflexive and separable Banach space continuously and densely embedded in a Hilbert space H. The scalar product in H is denoted by $(.;.)_H$ and the bilinear form on $V^* \times V$ is denoted by $(.;.)$. Let X be the space $L_2(S; V)$ where S denotes the time interval $[0, T]$. Therefore, we have $X^* = L_2(S; V^*)$ such that the bilinear form on $X^* \times X$ is given by

$$\langle x^*; x \rangle = \int_S (x^*(t); x(t))\, dt\,, \qquad x^* \in X^*, x \in X\,.$$

Let the abstract Sobolew space W consist of all $x \in X$ with a derivative $x' \in X^*$ in the sense of V^*-valued distributions. We remark that W is continuously embedded in $\mathbb{C}(S; H)$. Let U denote a topological vector space and $U_0 \subseteq U$ denotes the set of admissible controls.

The optimal control problem is the following one:

$$\left.\begin{aligned} &F(x, u) = \min!\\ &x' + A(x, u) = 0\,, \qquad x(0) = c(u)\,, \qquad H(x, u) \in Z_0\,, \qquad x \in W, u \in U_0\,, \end{aligned}\right\} \tag{9}$$

where $F\colon W \times U \to \mathbb{R}$, $A\colon X \times U \to X^*$, $c\colon U \to H$, $H\colon W \times U \to Z$, Z is a given Banach space and $Z_0 \subseteq Z$.

Let (x_0, u_0) be optimal for problem (9). We formulate the following assumptions of differentiability:

(A1) F and H have Gâteaux derivatives $F_x(x, u)$ and $H_x(x, u)$ with respect to $x \in W$ for all (x, u) of a neighbourhood of (x_0, u_0) and have also Gâteaux derivatives $F_u(x_0, u_0)$ and $H_x(x_0, u_0)$ with respect to $u \in U$ at the point (x_0, u_0).

F_x and H_x are pointwice continuous at (x_0, u_0), i.e. if $x_n \to x_0$ in W and $u_n \to u_0$ then $F_x(x_n, u_n)\, x \to F_x(x_0, u_0)\, x$ and $H_x(x_n, u_n)\, x \to H_x(x_0, u_0)\, x$ for each $x \in W$.

(A2) A has a Gâteaux derivative $A_x(x, u)$ with respect to $x \in X$ for all (x, u) of a neighbourhood of (x_0, u_0) and has also a Gâteaux derivative $A_u(x_0, u_0)$ with respect to $u \in U$ at the point (x_0, u_0). A_x is pointwice continuous at (x_0, u_0), i.e. if $x_n \to x_0$ in X and $u_n \to u_0$ then $A_x(x_n, u_n)\, x \to A_x(x_0, u_0)\, x$ for each $x \in X$.

(A3) c has a Gâteaux derivative $c_u(u_0)$ at the point u_0.

Defining the process equation $G(x, u) = 0$ by the evolution equation $x' + A(x, u) = 0$ including the initial condition $x(0) = c(u)$, i.e. defining $G\colon W \times U \to X^* \times H$ by

$$G(x, u) = (x' + A(x, u), x(0) - c(u))\,, \qquad x \in W, u \in U\,, \tag{10}$$

we see that all sequences $\{u_k\} \subseteq U_0$ which tend to u_0 are admissible sequences of varied controls for (x_0, u_0).

Further, let us assume that U_0 is a convex set. Then, considering $\{\gamma_k\} \subseteq \mathbb{R}_+$ with $\gamma_k \downarrow 0$ and $\{u_k\}$ with $u_k = u_0 + \gamma_k(u - u_0)$, $u \in U_0$, i.e. varied controls of Lagrange typ, special directional limits are

$$\Delta F = F_u(x_0, u_0)\,(u - u_0)\,, \qquad \Delta G = (A_u(x_0, u_0)\,(u - u_0), -c_u(u_0)\,(u - u_0))\,,$$
$$\Delta H = H_u(x_0, u_0)\,(u - u_0)\,.$$

The set Φ_0 of all such common directional limits $(\Delta F, \Delta G, \Delta H)$ is a convex cone.

Let us now consider the process equation $G(x, u) = 0$ to realize the assumptions (2.1)–(2.3). The essential basis is a general existence and uniqueness theorem for nonlinear evolution equations (see GAJEWSKI, GRÖGER and ZACHARIAS [4], ch. VI).

Definition: $A\colon X \to X^*$ is called *hemicontinuous* if the real function $s \to \langle A(x + sy); y \rangle$ is continuous for all $x, y \in X$.

A is called a *Volterra operator* if $x(s) = y(s)$ a.e. on $[0, t]$ implies always $A(x)(s) = A(y)(s)$ a.e. on $[0, t]$.

A is called *generalized strongly monotone* if there exist $m > 0$ and $\mu \geqq 0$ such that $\langle A(x) - A(y); x - y \rangle + \mu \|x - y\|^2_{L_2(S;H)} \geqq m \|x - y\|^2_X$ for all $x, y \in X$.

If $\mu = 0$ then A is strongly monotone in the usual sense.

Lemma 1: *Let $A\colon X \to X^*$ be a hemicontinuous Volterra operator which is generalized strongly monotone. Let $f^* \in X^*$ and $c \in H$. Then the evolution equation $x' + A(x) = f^*$, $x(0) \in c$, $x \in W$, has a unique solution.*

The realization of the assumptions (2.1)–(2.3) is the following one.

Theorem 3: *Let the assumptions* (A2) *and* (A3) *be fulfilled, and suppose*

(A4) *$A(., u)\colon X \to X^*$ is a hemicontinuous Volterra operator for all u of a neighbourhood of u_0 which is generalized strongly monotone with constants $m > 0$ and $\mu \geqq 0$ independent of u.*

Let $\{\gamma_k\} \subseteq \mathbb{R}_+$ with $\gamma_k \downarrow 0$ and $\{u_k\}$ with $u_k = u_0 + \gamma_k(u - u_0)$, $u \in U_0$. Then the assumptions (2.1)–(2.3) *are fulfilled, i.e.:*

$$\textit{For all } k \gg 1 \textit{ there exists } x_k \in W \textit{ such that } x_k' + A(x_k, u_k) = 0\,, \qquad x_k(0) = c(u_k)\,. \tag{11.1}$$

$$\Delta x = \lim_{k \to \infty} \frac{1}{\gamma_k}\,(x_k - x_0)\ \textit{exists}\,. \tag{11.2}$$

$$\left.\begin{array}{l}\Delta x \text{ is the unique solution of the variational evolution equation} \\ \Delta x' + A_x(x_0, u_0)\,\Delta x = -A_u(x_0, u_0)\,(u - u_0), \qquad \Delta x(0) = c_u(u_0)\,(u - u_0).\end{array}\right\} \tag{11.3}$$

Proof: Assumption (A4) implies that $A_x(x, u)\colon X \to X^*$ is generalized strongly monotone with the constants $m > 0$ and $\mu \geqq 0$ for all $x \in X$ and all u in a neighbourhood of u_0. Using lemma 1 (11.1) and (11.3) are valid. We have only to show that $\Delta x = \lim\limits_{k\to\infty} \frac{1}{\gamma_k}(x_k - x_0)$.

First, we assume $\mu = 0$, i.e. $A(., u)\colon X \to X^*$ is strongly monotone in the usual sense.

Remembering that for $x, y \in W$ and $s, t \in S$ the following rule of integration by parts is valid,

$$(x(t), y(t))_H - (x(s); y(s))_H = \int_s^t [(x'(\tau); y(\tau)) + (y'(\tau); x(\tau))]\,d\tau,$$

and applying this rule of integration by parts we derive

$$\begin{aligned}&\tfrac{1}{2}\|x_k(T) - x_0(T)\|_H^2 - \tfrac{1}{2}\|x_k(0) - x_0(0)\|_H^2 \\ &= \langle x_k' - x_0'; x_k - x_0\rangle = \langle -A(x_k, u_k) + A(x_0, u_0); x_k - x_0\rangle,\end{aligned}$$

and further,

$$\begin{aligned}&\tfrac{1}{2}\|x_k(T) - x_0(T)\|_H^2 + \langle A(x_k, u_k) - A(x_0, u_k); x_k - x_0\rangle \\ &= \langle A(x_0, u_0) - A(x_0, u_k); x_k - x_0\rangle + \tfrac{1}{2}\|x_k(0) - x_0(0)\|_H^2,\end{aligned}$$

and using the monotonicity and the initial conditions,

$$m\|x_k - x_0\|_X^2 \leqq \|A(x_0, u_k) - A(x_0, u_0)\|_{X^*}\,\|x_k - x_0\|_X + \tfrac{1}{2}\|c(u_k) - c(u_0)\|_H^2,$$

and dividing by γ_k^2,

$$m\left\|\frac{1}{\gamma_k}(x_k - x_0)\right\|_X^2 \leqq \left\|\frac{1}{\gamma_k}(A(x_0, u_k) - A(x_0, u_0))\right\|_{X^*}\left\|\frac{1}{\gamma_k}(x_k - x_0)\right\|_X + \frac{1}{2}\left\|\frac{1}{\gamma_k}(c(u_k) - c(u_0))\right\|_H^2.$$

It follows that $\left\{\frac{1}{\gamma_k}(x_k - x_0)\right\}$ is bounded in X.

Oncemore, applying integration by parts we obtain

$$\begin{aligned}&\frac{1}{2}\left\|\frac{1}{\gamma_k}(x_k(T) - x_0(T)) - \Delta x(T)\right\|_H^2 - \frac{1}{2}\left\|\frac{1}{\gamma_k}(x_k(0) - x_0(0)) - \Delta x(0)\right\|_H^2 \\ &= \left\langle \frac{1}{\gamma_k}(x_k' - x_0') - \Delta x'; \frac{1}{\gamma_k}(x_k - x_0) - \Delta x\right\rangle \\ &= \left\langle \frac{1}{\gamma_k}(-A(x_k, u_k) + A(x_0, u_0)) + A_x(x_0, u_0)\,\Delta x + A_u(x_0, u_0)\,(u - u_0); \frac{1}{\gamma_k}(x_k - x_0) - \Delta x\right\rangle,\end{aligned}$$

and further,

$$\begin{aligned}&\frac{1}{2}\left\|\frac{1}{\gamma_k}(x_k(T) - x_0(T)) - \Delta x(T)\right\|_H^2 + \left\langle \frac{1}{\gamma_k}(A(x_k, u_k) - A(x_0, u_k)) - A_x(x_0, u_0)\,\Delta x; \frac{1}{\gamma_k}(x_k - x_0) - \Delta x\right\rangle \\ &= \left\langle -\frac{1}{\gamma_k}(A(x_0, u_k) - A(x_0, u_0)) + A_u(x_0, u_0)\,(u - u_0); \frac{1}{\gamma_k}(x_k - x_0) - \Delta x\right\rangle \\ &\quad + \frac{1}{2}\left\|\frac{1}{\gamma_k}(x_k(0) - x_0(0)) - \Delta x(0)\right\|_H^2,\end{aligned}$$

and using the usual mean value theorem and the initial conditions,

$$\begin{aligned}&\left\langle A_x(x_0 + \vartheta_k(x_k - x_0), u_k)\left(\frac{1}{\gamma_k}(x_k - x_0) - \Delta x\right); \frac{1}{\gamma_k}(x_k - x_0) - \Delta x\right\rangle \\ &+ \left\langle (A_x(x_0 + \vartheta_k(x_k - x_0), u_k) - A_x(x_0, u_0))\,\Delta x; \frac{1}{\gamma_k}(x_k - x_0) - \Delta x\right\rangle \\ &\leqq \left\langle -\frac{1}{\gamma_k}(A(x_0, u_k) - A(x_0, u_0)) + A_u(x_0, u_0)\,(u - u_0); \frac{1}{\gamma_k}(x_k - x_0) - \Delta x\right\rangle \\ &\quad + \frac{1}{2}\left\|\frac{1}{\gamma_k}(c(u_k) - c(u_0)) - c_u(u_0)\,(u - u_0)\right\|_H^2,\end{aligned}$$

and using the monotonicity,

$$\begin{aligned} m \left\| \frac{1}{\gamma_k} (x_k - x_0) - \Delta x \right\|_X^2 &\leqq \|(A_x(x_0 + \vartheta_k(x_k - x_0), u_k) - A_x(x_0, u_0)) \Delta x\|_{X^*} \left\| \frac{1}{\gamma_k} (x_k - x_0) - \Delta x \right\|_X \\ &+ \left\| \frac{1}{\gamma_k} (A(x_0, u_k) - A(x_0, u_0)) - A_u(x_0, u_0)(u - u_0) \right\|_{X^*} \left\| \frac{1}{\gamma_k} (x_k - x_0) - \Delta x \right\|_X \\ &+ \frac{1}{2} \left\| \frac{1}{\gamma_k} (c(u_k) - c(u_0)) - c_u(u_0)(u - u_0) \right\|_H^2 . \end{aligned}$$

It follows that $\left\{\frac{1}{\gamma_k} (x_k - x_0)\right\}$ tends to Δx in X.

If we show that $\left\{\frac{1}{\gamma_k} (x_k' - x_0')\right\}$ tends to $\Delta x'$ in X^*, too, then $\left\{\frac{1}{\gamma_k} (x_k - x_0)\right\}$ tends to Δx in W, and the proof is complete.

We get this convergence from the following calculation:

$$\begin{aligned} \left\| \frac{1}{\gamma_k} (x_k' - x_0') - \Delta x' \right\|_{X^*} &= \left\| \frac{1}{\gamma_k} (A(x_k, u_k) - A(x_0, u_0)) - A_x(x_0, u_0) \Delta x - A_u(x_0, u_0)(u - u_0) \right\|_{X^*} \\ &\leqq \left\| \frac{1}{\gamma_k} (A(x_k, u_k) - A(x_0, u_k)) - A_x(x_0, u_0) \Delta x \right\|_{X^*} \\ &+ \left\| \frac{1}{\gamma_k} (A(x_0, u_k) - A(x_0, u_0)) - A_u(x_0, u_0)(u - u_0) \right\|_{X^*} \\ &\leqq \left\| A_x(x_0 + \vartheta_k(x_k - x_0), u_k) \frac{1}{\gamma_k} (x_k - x_0) - A_x(x_0, u_0) \Delta x \right\|_{X^*} \\ &+ \left\| \frac{1}{\gamma_k} (A(x_0, u_k) - A(x_0, u_0)) - A_u(x_0, u_0)(u - u_0) \right\|_{X^*} \\ &\leqq \|A_x(x_0 + \vartheta_k(x_k - x_0), u_k)\|_{\mathscr{L}(X;X^*)} \left\| \frac{1}{\gamma_k} (x_k - x_0) - \Delta x \right\|_X \\ &+ \|(A_x(x_0 + \vartheta_k(x_k - x_0), u_k) - A_x(x_0, u_0)) \Delta x\|_{X^*} \\ &+ \left\| \frac{1}{\gamma_k} (A(x_0, u_k) - A(x_0, u_0)) - A_u(x_0, u_0)(u - u_0) \right\|_{X^*} . \end{aligned}$$

Finally, we assume $\mu > 0$. We remark only that we consider instead of the evolution equation

$$x' + A(x, u) = 0\,, \qquad x(0) = c(u)\,, \qquad x \in W\,,$$

the equivalent evolution equation

$$\bar{x}' + \bar{A}(\bar{x}, u) = 0\,, \qquad \bar{x}(0) = c(u)\,, \qquad \bar{x} \in W\,,$$

where $\bar{A}: X \times U \to X^*$ is given by

$$\bar{A}(\bar{x}, u)(t) = e^{-\mu t} A(e^{\mu} \cdot \bar{x}, u)(t) + \mu I x(t)\,, \qquad t \in S\,,$$

and $I: V \to V^*$ is defined by the embedding $V \subseteq H \subseteq V^*$.

The connexion is the following. If $x \in W$ is a solution of the first evolution equation then $\bar{x} = e^{-\mu} \cdot x$ is a solution of the second one, and conversely. But, the operator $\bar{A}(., u): X \to X^*$ is strongly monotone in the usual sense (compare Gajewski, Gröger and Zacharias [4], ch. VI). ■

2.2 Necessary optimality conditions – int Z_0 nonempty

Theorem 4: *Let the assumptions* (A1)–(A4) *be fulfilled. Let $U_0 \subseteq U$ be a convex set and let $Z_0 \subseteq Z$ be a convex set with a nonempty interior. Then a necessary optimality condition for (x_0, u_0) to be optimal for problem* (9) *is the following one:*

There exist $\lambda \geqq 0$ and $z^* \in Z^*$ (λ and z^* not both zero) such that there hold:

the variational inequality

$$\begin{aligned} &\langle z^*; H_x(x_0, u_0) \Delta x + H_u(x_0, u_0)(u - u_0)\rangle \\ &+ \lambda(F_x(x_0, u_0) \Delta x + F_u(x_0, u_0)(u - u_0)) \geqq 0 \quad \textit{for all} \quad u \in U_0\,, \end{aligned} \tag{12.1}$$

where Δx is the unique solution of the variational evolution equation

$$\Delta x' + A_x(x_0, u_0)\, \Delta x = -A_u(x_0, u_0)\,(u - u_0), \qquad \Delta x(0) = c_u(u_0)\,(u - u_0), \qquad \Delta x \in W, \tag{12.2}$$

and

the maximum condition

$$\langle z^*; H(x_0, u_0)\rangle = \max_{z \in Z_0} \langle z^*; z\rangle. \tag{12.3}$$

Proof: The assertion follows from theorems 2 and 3. ■

Considering a corresponding adjoint evolution we obtain the following corollary.

Corollary: *The variational inequality* (12.1) *can be written as follows:*

$$\left.\begin{aligned}&\langle z^*; H_u(x_0, u_0)\,(u - u_0)\rangle - \langle A_u(x_0, u_0)\,(u - u_0); y + \lambda f + h\rangle\\ &+ (c_u(u_0)\,(u - u_0); y(0))_H + \lambda F_u(x_0, u_0)\,(u - u_0) \geqq 0 \quad \textit{for all} \quad u \in U_0,\end{aligned}\right\} \tag{13.1}$$

where y is the unique solution of the adjoint evolution equation

$$-y' + A_x^*(x_0, u_0)\, y = \lambda f^* + h^* - A_x^*(x_0, u_0)\,(\lambda f + h), \qquad y(T) = 0, \qquad y \in W, \tag{13.2}$$

and (f^*, f), $(h^*, h) \in X^* \times X$ are arbitrary representations of $F_x(x_0, u_0)$ and $H_x^*(x_0, u_0)\, z^*$.

(Each $w^* \in W^*$ can be represented by $(l^*, l) \in X^* \times X$ such that $\langle w^*; w\rangle = \langle l^*; w\rangle + \langle w'; l\rangle$ for all $w \in W$.)

Proof: Applying the corollary of theorem 2 we have to consider the general adjoint equation (7) and the special adjoint equation (8.1). In our case (see (10)) $G_x(x_0, u_0)$: $W \to X^* \times H$ is given by

$$G_x(x_0, u_0)\, w = (w' + A_x(x_0, u_0)\, w, w(0)), \qquad w \in W.$$

Therefore, the adjoint operator $G_x^*(x_0, u_0)$: $X \times H \to W^*$ is defined by the formula

$$\begin{aligned}\langle G_x^*(x_0, u_0)\,(x, h); w\rangle &= \langle G_x(x_0, u_0)\, w; (x, h)\rangle\\ &= \langle w' + A_x(x_0, u_0)\, w; x\rangle + (w(0); h)_H, \qquad x \in X, h \in H, w \in W.\end{aligned}$$

Now the adjoint equation (7) is the following one:

$$\langle w' + A_x(x_0, u_0)\, w; x\rangle + (w(0); h)_H = \langle l^*; w\rangle + \langle w'; l\rangle \quad \text{for all} \quad w \in W,$$

where $(l^*, l) \in X^* \times X$ is a representation of the given right hand side $w^* \in W^*$. It follows

$$\langle w'; x - l\rangle + \langle A_x^*(x_0, u_0)\, x - l^*; w\rangle + (w(0); h)_H = 0 \quad \text{for all} \quad w \in W.$$

We set $w(t) = \varphi(t)\, v$, $t \in S$, where $\varphi \in \mathbb{C}_0^\infty(\text{int } S)$ and $v \in V$, which implies $y = x - l \in W$ and $y' = A_x^*(x_0, u_0)\, x - l^*$, i.e.

$$-y' + A_x^*(x_0, u_0)\, y = l^* - A_x^*(x_0, u_0)\, l, \qquad y \in W. \tag{14.1}$$

Suppose that y is a solution of (14.1). Then we get

$$\langle w'; y\rangle + \langle y'; w\rangle + (w(0); h)_H = 0 \quad \text{for all} \quad w \in W,$$

and using integration by parts,

$$(w(T); y(T))_H + (w(0); h - y(0))_H = 0 \quad \text{for all} \quad w \in W.$$

Considering $w(t) = (T - t)\, v$, $w(t) = tv$, $t \in S$, $v \in V$, we get $y(0) = h$ and

$$y(T) = 0. \tag{14.2}$$

That means, if $(x, h) \in X \times H$ is a solution of the adjoint equation $G_x^*(x_0, u_0)\,(x, h) = w^*$ then $y = x - l \in W$ is a solution of the adjoint evolution equation (14.1) including (14.2), and conversely, if $y \in W$ is a solution of the adjoint evolution equation (14.1) with (14.2) then $(x, h) \in X \times H$ given by $x = y + l$ and $h = y(0)$ is a solution of the adjoint equation $G_x^*(x_0, u_0)\,(x, h) = w^*$.

Obviously, $A_x^*(x_0, u_0)$: $X \to X^*$ is generalized strongly monotone. A similar existence and uniqueness theorem (see lemma 1) for an evolution equation backward in time can be formulated, i.e. the adjoint evolution equation (14.1) including (14.2) has a unique solution.

The special adjoint equation (8.1) contains the right hand side $w^* = \lambda F_x(x_0, u_0) + H_x^*(x_0, u_0)\, z^*$, which implies $l^* = \lambda f^* + h^*$ and $l = \lambda f + h$. ■

2.3. Necessary optimality conditions – int Z_0 empty

The assumption of a nonempty interior of Z_0 is very strong. For instance, the set

$$Z_0 = \{z \in L_2(S) \mid z_1(t) \leqq z(t) \leqq z_2(t) \text{ a.e.}\}, \qquad z_1, z_2 \in L_2(S),$$

has no interior point in $Z = L_2(S)$.

It is our aim to save the necessary optimality conditions for a set Z_0 with an empty interior.

The idea is the following one. Let (x_0, u_0) be an optimal solution of the control problem (9) where the space of the controls U is a Hilbert space and the Banach space Z is reflexive. Then we consider the approximated control problems

$$\left.\begin{aligned} &F(x, u) + \frac{\alpha}{2} \|u - u_0\|_U^2 = \min! \\ &x' + A(x, u) = 0, \quad x(0) = c(u), \quad H(x, u) \in Z_n, \quad x \in W, u \in U_0, \end{aligned}\right\} \tag{15}$$

where $Z_n = \bigcup_{z \in Z_0} \left\{z' \in Z \mid \|z' - z\| \leqq \frac{1}{n}\right\}$, $n \in \mathbb{N}$, and $\alpha > 0$ will be fixed later.

If (x_n, u_n) is an optimal solution of the control problem (15) then we are able to formulate necessary optimality conditions using theorem 4. We will derive necessary optimality conditions also for (x_0, u_0) from the convergence $x_n \to x_0$ and $u_n \to u_0$ as $n \to \infty$.

We formulate similar, but more general assumptions than (A1)–(A4) where the set $U_0' \subseteq U_0$ will be fixed later.

(B1) F: $W \times U \to \mathbb{R}$ and H: $W \times U \to \mathbb{R}$ are Gâteaux differentiable with respect to $x \in W$ and $u \in U$ on $W \times U_0'$ and the Gâteaux derivatives F_x, F_u, H_x and H_u are pointwice continuous on $W \times U_0'$.

(B2) A: $X \times U \to X^*$ is Gâteaux differentiable with respect to $x \in X$ and $u \in U$ on $X \times U_0'$ and the Gâteaux derivatives A_x and A_u are pointwice continuous on $X \times U_0'$.

(B3) c: $U \to H$ is Gâteaux differentiable on U_0' and the Gâteaux derivative c_u is pointwice continuous on U_0'.

(B4) $A(., u)$: $X \to X^*$ is a Volterra operator for all $u \in U_0'$ which is generalized strongly monotone with constants $m > 0$ and $\mu \geqq 0$ independent of $u \in U_0'$.

Let χ be a topology of U which is weaker than or equal to the norm topology. Let τ be the topology of W given by the norm convergence in X, by the weak convergence of the derivatives in X^* and by the norm convergence in $\mathbb{C}(S; H)$. σ denotes always the weak topology.

An existence theorem for the control problem (15) is the following one.

Lemma 2: *Let N be a closed ball with center u_0 such that $U_0' = U_0 \cap N$ is χ-compact and let Z_n be weakly compact. Suppose that the assumption* (B4) *is fulfilled and that*:

(B5) $A(x, .)$: $(U, \chi) \to (X^*, \sigma)$, $A(x, .)$: $(U, \chi) \to W^*$, c: $(U, \chi) \to H$ are continuous on U_0'.

(B6) H: $(W, \tau) \times (U, \chi) \to (Z, \sigma)$ is continuous on $W \times U_0'$.

(B7) F: $(W, \tau) \times (U, \chi) \to \mathbb{R}$, $\|\cdot\|_U$: $(U, \chi) \to \mathbb{R}$ are lower semicontinuous and F has a lower bound F_{inf}.

Then for sufficiently large $\alpha > 0$ the control problem (15) *has an optimal solution (x_n, u_n) where $u_n \in U_0'$ holds.*

Proof: We can consider the problem (15) with U_0' instead of U_0 because

$$F_{\text{inf}} + \frac{\alpha}{2} \|u_n - u_0\|_U^2 \leqq F(x_n, u_n) + \frac{\alpha}{2} \|u_n - u_0\|_U^2 \leqq F(x_0, u_0),$$

and therefore, $u_n \in U_0' = U_0 \cap N$ for sufficiently large $\alpha > 0$.

The set $\Psi = \{(x, u) \in W \times U_0' \mid x' + A(x, u) = 0, x(0) = c(u)\}$ is (τ, χ)-compact (see [10], or [12], ch. 2). The existence follows from the wellknown Weierstrass theorem. ■

Remark: Suppose that the topology χ is given by the norm convergence. Then the assumption (B2) implies the assumption (B5).

We get a necessary optimality condition for (x_0, u_0) as follows.

Theorem 5: *Let $U_0 \subseteq U$ be convex, let N be a closed ball with center u_0 such that $U_0' = U_0 \cap N$ is χ-compact and let $Z_0 \subseteq Z$ be convex and closed. Suppose that the assumptions* (B1)–(B7) *are fulfilled.*

Then the necessary optimality conditions of theorem 4 and of the corollary for (x_0, u_0) to be optimal for problem (9) *holds, but the nontriviality of $\lambda \geqq 0$ and $z^* \in Z^*$ is uncertain.*

Proof: Applying lemma 2 for each $n \in \mathbb{N}$ the control problem (15) has an optimal solution (x_n, u_n). Because $(x_n, u_n) \in \Psi$ and Ψ is (τ, χ)-compact (see proof of lemma 2) we can assume that

$$x_n \to x \quad \text{in } (W, \tau) \quad \text{and} \quad u_n \to u \quad \text{in } (U, \chi) \quad \text{as } n \to \infty .$$

Furthermore, $H(x, u) \in Z_0$, i.e. (x, u) is an admissible solution of the control problem (9).

The following inequality holds:

$$F(x_0, u_0) \leqq F(x, u) \leqq F(x, u) + \frac{\alpha}{2} \|u - u_0\|_U^2 \leqq \liminf_{n \to \infty} \left(F(x_n, u_n) + \frac{\alpha}{2} \|u_n - u_0\|_U^2 \right) \leqq F(x_0, u_0),$$

which implies $u = u_0$ and $x = x_0$.

$\|x'_n - x'_0\|_{X^*} = \|A(x_n, u_n) - A(x_0, u_0)\|_{X^*}$ and the continuity of A imply $\|x'_n - x'_0\|_{X^*} \to 0$ and $\|x_n - x_0\|_W \to 0$. Indeed, $x_n \to x_0$ in W and $u_n \to u_0$ in U.

Applying theorem 4 the following necessary optimality condition for (x_n, u_n) holds, i.e. there exists $\lambda_n \geqq 0$ and $z_n^* \in Z^*$ (λ_n and z_n^* not both zero) such that

$$\langle z_n^*; H_x(x_n, u_n)\, \Delta_n x + H_u(x_n, u_n)\,(u - u_n)\rangle + \lambda_n\{F_x(x_n, u_n)\, \Delta_n x + F_u(x_n, u_n)\,(u - u_n)\}$$

$$+ \alpha(u_n - u_0; u - u_n)_U \geqq 0 \qquad \text{for all } u \in U_0, \tag{16.1}$$

$$\Delta_n x' + A_x(x_n, u_n)\, \Delta_n x = -A_u(x_n, u_n)\,(u - u_n), \qquad \Delta_n x(0) = c_u(u_n)\,(u - u_n), \qquad \Delta_n x \in W, \tag{16.2}$$

$$\langle z_n^*, H(x_n, u_n)\rangle = \max_{z \in Z_n} \langle z_n^*, z\rangle. \tag{16.3}$$

We can assume that $\lambda_n + \|z_n^*\|_{Z^*} = 1$ and therefore,

$$\lambda_n \to \lambda \quad \text{and} \quad z_n^* \to z^* \qquad \text{in } (Z^*, \sigma).$$

Using the assumptions (B1)–(B3) the convergence $n \to \infty$ in the conditions (16.1)–(16.3) for (x_n, u_n) to the conditions (12.1)–(12.3) for (x_0, u_0) is valid.

We have only to show that $\Delta_n x \to \Delta x$ in W. To be short we write the variational evolution equations (16.2) and (12.2) as follows:

$$\Delta_n x' + L_n\, \Delta_n x = f_n, \qquad \Delta_n x(0) = a_n, \qquad \Delta x' + L\, \Delta x = f, \qquad \Delta x(0) = a.$$

First, we assume $\mu = 0$, i.e. L_n, $L: X \to X^*$ are strongly monotone in the usual sense.

Applying integration by parts we obtain

$$\tfrac{1}{2} \|\Delta_n x(T) - \Delta x(T)\|_H^2 - \tfrac{1}{2} \|\Delta_n x(0) - \Delta x(0)\|_H^2 = \langle \Delta_n x' - \Delta x'; \Delta_n x - \Delta x\rangle$$

$$= \langle f_n - f; \Delta_n x - \Delta x\rangle - \langle L_n\, \Delta_n x - L\, \Delta x; \Delta_n x - \Delta x\rangle,$$

and further,

$$\langle L_n(\Delta_n x - \Delta x); \Delta_n x - \Delta x\rangle \leqq \tfrac{1}{2} \|a_n - a\|_H^2 + \langle f_n - f; \Delta_n x - \Delta x\rangle - \langle (L_n - L)\, \Delta x; \Delta_n x - \Delta x\rangle,$$

and using the monotonicity,

$$m\, \|\Delta_n x - \Delta x\|_X^2 \leqq \tfrac{1}{2} \|a_n - a\|_H^2 + (\|f_n - f\|_{X^*} + \|(L_n - L)\, \Delta x\|_{X^*})\, \|\Delta_n x - \Delta x\|_X.$$

It follows that $\{\|\Delta_n x - \Delta x\|_X\}$ is bounded, and therefore, $\Delta_n x \to \Delta x$ in X follows from $a_n \to a$ in H, $f_n \to f$ in X^* and $L_n \to L$ pointwise.

The estimation

$$\|\Delta_n x' - \Delta x'\|_{X^*} \leqq \|L_n\, \Delta_n x - L\, \Delta x\|_{X^*} + \|f_n - f\|_{X^*}$$

$$\leqq \|L_n\|_{\mathscr{L}(X; X^*)}\, \|\Delta_n x - \Delta x\|_X + \|(L_n - L)\, \Delta x\|_{X^*} + \|f_n - f\|_{X^*}$$

implies $\Delta_n x' \to \Delta x'$ in X^*, and finally, $\Delta_n x \to \Delta x$ in W.

For the case $\mu > 0$ we consider the equivalent evolution equations

$$\overline{\Delta_n x'} + \bar{L}_n\, \overline{\Delta_n x} = \bar{f}_n, \qquad \overline{\Delta_n x}(0) = a_n, \qquad \overline{\Delta x'} + \bar{L}\, \overline{\Delta x} = \bar{f}, \qquad \overline{\Delta x}(0) = a,$$

where

$$\overline{L_n x}(t) = \mathrm{e}^{-\mu t}\, (L_n\, \mathrm{e}^{\mu \cdot} \cdot \bar{x})\, (t) + \mu I \bar{x}(t), \qquad \bar{f}_n(t) = \mathrm{e}^{-\mu t}\, f_n(t),$$

$$\overline{Lx}(t) = \mathrm{e}^{-\mu t}\, (L\, \mathrm{e}^{\mu \cdot} \cdot \bar{x})\, (t) + \mu I \bar{x}(t), \qquad \bar{f}(t) = \mathrm{e}^{-\mu t}\, f(t), \qquad t \in S.$$

(Compare the proof of theorem 3.) ■

3. Examples

Let Ω be a bounded domain of $\mathbb{R}^n$. We consider optimal control problems governed by nonlinear initial-boundary value problems for parabolic differential equations of second order of divergence typ.

3.1. Optimal control in Dirichlet boundary conditions

$$\left.\begin{aligned}
&\frac{1}{2}\int\limits_S\int\limits_\Omega |x(t,s)-\bar{x}(t,s)|^2\,ds\,dt+\frac{1}{2}\int\limits_S\int\limits_{\partial\Omega}|u(t,s)|^2\,ds\,dt=\min!\\
&\frac{\partial x}{\partial t}(t,s)-\sum_{i=1}^{n}\frac{\partial}{\partial s_i}a_i\left(t,s,\frac{\partial x}{\partial s_i}(t,s)\right)+a_0(t,s,x(t,s))=0\,,\qquad t\in S,\ s\in\Omega\,,\\
&x(0,s)=c(s)\,,\qquad s\in\Omega\,,\qquad x(t,s)=u(t,s)\,,\qquad t\in S,\ s\in\partial\Omega\,,\\
&u_1(t,s)\leqq u(t,s)\leqq u_2(t,s)\qquad \text{a.e. on } S\times\partial\Omega\,.
\end{aligned}\right\}\tag{17.1}$$

To solve the initial-boundary value problem we need a regularity condition for the given boundary value (control) function $u(.)$ (see [12], [7]). Therefore, the necessary optimality conditions are very complicated. In this paper we choose a new way. We introduce an artificial Neumann boundary condition

$$\frac{\partial x}{\partial v_s}(t,s)=w(t,s)\qquad \text{a.e. on } S\times\partial\Omega\tag{17.2}$$

where $\frac{\partial x}{\partial v_s}(t,s)$ denotes the derivative in direction of the outer conormal

$$\frac{\partial x}{\partial v_s}(t,s)=\sum_{i=1}^{n}a_i\left(t,s,\frac{\partial x}{\partial s_i}(t,s)\right)\cos(v_s;s_i)$$

(v_s denotes the unit vector of the outer normal in $s\in\partial\Omega$).

The introduced control function $w(.)$ is not subject to any condition. The Dirichlet boundary condition with the real control function $u(.)$ is considered as an additional state condition for $x(.)$.

To formulate the control problem (17.1), (17.2) as a problem (9) we set

$$V=W_2^{(1)}(\Omega)\,,\qquad H=L_2(\Omega)\,,\qquad U_0=U=L_2(S\times\partial\Omega)\,,$$

$$F(x,w)=\frac{1}{2}\int\limits_S\int\limits_\Omega|x(t,s)-\bar{x}(t,s)|^2\,ds\,dt+\frac{1}{2}\int\limits_S\int\limits_{\partial\Omega}|x(t,s)|^2\,ds\,dt\,,$$

$$\langle A(x,w)(t),v\rangle=\int\limits_\Omega\left\{\sum_{i=1}^{n}a_i\left(t,s,\frac{\partial}{\partial s_i}(t,s)\right)\frac{\partial v}{\partial s_i}(s)+a_0(t,s,x(t,s))\,v(s)\right\}ds$$

$$-\int\limits_{\partial\Omega}w(t,s)\,v(s)\,ds\,,\qquad t\in S,\ x\in X,\ w\in U,\ v\in V\,,$$

$$Z=L_2(S\times\partial\Omega)\,,\qquad Z_0=\{u\in Z\mid u_1(t,s)\leqq u(t,s)\leqq u_2(t,s)\ \text{a.e. on } S\times\partial\Omega\}\,,$$

$$H(x,u)(t,s)=x(t,s)|_{\partial\Omega}\,,\qquad t\in S,\ x\in W,\ w\in U\,.$$

Remark: The formulation is correct if the following assumptions are fulfilled:

a) $a_i(t,s,x)$, $i=0,1,\ldots,n$, are measurable with respect to t, s and continuous with respect to x.

b) For all $i=0,1,\ldots,n$ there exist $\gamma_i\in L_2(S\times\Omega)$ and $\beta_i\geqq 0$ such that

$$|a_i(t,s,x)|\leqq\gamma_i(t,s)+\beta_i|x|\qquad \text{a.e. on } S\times\Omega\,,\quad \text{for all}\quad x\in\mathbb{R}\,.$$

c) $\bar{x}\in L_2(S\times\Omega)$, $\quad c\in L_2(\Omega)$, $\quad u_1,u_2\in L_2(S\times\partial\Omega)$.

To derive necessary optimality conditions we assume further that

d) $a_i(t,s,x)$, $i=0,1,\ldots,n$, are continuously differentiable with respect to x.

e) For all $i=0,1,\ldots,n$ there exist $M_i\geqq m_i>0$ such that

$$m_i\leqq\frac{\partial}{\partial x}a_i(t,s,x)\leqq M_i\qquad \text{a.e. on } S\times\Omega \text{ for all } x\in\mathbb{R}\,.$$

Let us suppose that $(x_0,w_0)\in W\times U_0$ is optimal for problem (17.1), (17.2) in the formulation as problem (9). Then we obtain from theorem 4 and the corollary:

There exist $\lambda\geqq 0$ and $z^*\in Z$ such that there holds the variational inequality

$$-\langle A_W(x_0,w_0)(w-w_0),y+\lambda f+h\rangle\geqq 0\qquad \text{for all } w\in U_0\,,$$

where y is the solution of the adjoint evolution equation

$$-y'+A_x^*(x_0,w_0)\,y=\lambda f^*+h^*-A_x^*(x_0,u_0)(\lambda f+h)\,,\qquad y(T)=0\,,\qquad y\in W\,,$$

and the maximum condition is valid:

$$\langle z^*;H(x_0,w_0)\rangle=\max_{z\in Z_0}\langle z^*;z\rangle\,.$$

We calculate where $t \in S$ and $v \in V$

$$(f^*(t); v) = \int_{\Omega} (x_0(t, s) - \bar{x}(t, s))\, v(s)\, ds + \int_{\partial\Omega} x_0(t, s)\, v(s)\, ds\,,$$

$$f(t) = 0\,, \qquad (h^*(t); v) = \int_{\partial\Omega} z^*(t, s)\, v(s)\, ds\,, \qquad h(t) = 0\,.$$

Therefore, we obtain:

There exist $\lambda \geqq 0$ and $z^* \in Z^*$ such that

$$y(t, s) = 0\,, \qquad t \in S\,,\ s \in \partial\Omega\,,$$

where

$$-\frac{\partial y}{\partial t}(t, s) - \sum_{i=1}^{n} \frac{\partial}{\partial s_i} \frac{\partial}{\partial x} a_i\left(t, s, \frac{\partial x_0}{\partial s_i}(t, s)\right) \frac{\partial y}{\partial s_i}(t, s)$$

$$+ \frac{\partial}{\partial x} a_0(t, s, x_0(t, s))\, y(t, s) = \lambda(x_0(t, s) - \bar{x}(t, s))\,, \qquad t \in S,\ s \in \Omega\,,$$

$$y(T, s) = 0\,, \qquad s \in \Omega\,, \qquad \frac{\partial y}{\partial v_s}(t, s) = \lambda x_0(t, s) + z^*(t, s)\,, \qquad t \in S,\ s \in \partial\Omega\,,$$

$$\left(\frac{\partial x}{\partial v_s}(t, s) = \sum_{i=1}^{n} \frac{\partial}{\partial x} a_i\left(t, s, \frac{\partial x_0}{\partial s_i}(t, s)\right) \frac{\partial y}{\partial s_i}(t, s) \cos(v_s, s_i)\right)$$

and

$$z^*(t, s)\, x_0(t, s) = \max_{u_1(t,s) \leqq u \leqq u_2(t,s)} z^*(t, s)\, u \qquad \text{a.e. on } S \times \partial\Omega\,,$$

or, there exists $\lambda \geqq 0$ such that

$$\left(\frac{\partial y}{\partial v_s}(t, s) - \lambda x_0(t, s)\right) x_0(t, s) = \max_{u_1(t,s) \leqq u \leqq u_2(t,s)} \left(\frac{\partial y}{\partial v_s}(t, s) - \lambda x_0(t, s)\right) u \qquad \text{a.e. on } S \times \partial\Omega\,,$$

where

$$-\frac{\partial y}{\partial t}(t, s) - \sum_{i=1}^{n} \frac{\partial}{\partial s_i} \frac{\partial}{\partial x} a_i\left(t, s, \frac{\partial x_0}{\partial s_i}(t, s)\right) \frac{\partial y}{\partial s_i}(t, s)$$

$$+ \frac{\partial}{\partial x} a_0(t, s, x_0(t, s))\, y(t, s) = \lambda(x_0(t, s) - \bar{x}(t, s))\,, \qquad t \in S,\ s \in \Omega\,,$$

$$y(T, s) = 0\,, \qquad s \in \Omega\,, \qquad y(t, s) = 0\,, \qquad t \in S,\ s \in \partial\Omega\,.$$

(y is the solution of the adjoint evolution equation belonging to this adjoint initial-boundary value problem which is only correct under additional regularity conditions for $a_i(.)$ and x_0.)

3.2. Time-optimal control

$$\left.\begin{array}{l}
T = \min! \\
\dfrac{\partial x}{\partial t}(t, s) - \displaystyle\sum_{i=1}^{n} \frac{\partial}{\partial s_i} a_i\left(t, s, \frac{\partial x}{\partial s_i}(t, s), d^i\right) \\
\qquad + a_0(t, s, x(t, s), d^0)\, u(t, s) = 0\,, \qquad t \in [0, T],\ s \in \Omega\,, \\
x(0, .) \in H_0\,, \qquad x(T, .) \in H_1\,, \qquad x(t, s) = 0\,, \qquad t \in [0, T],\ s \in \partial\Omega\,, \\
u_1(s) \leqq u(t, s) \leqq u_2(s) \qquad \text{a.e. on } [0, T] \times \Omega\,, \\
d^i \in [d_1^i, d_2^i]\,, \qquad i = 0, 1, \ldots, n\,, \qquad T \geqq 0\,.
\end{array}\right\} \tag{18}$$

To derive necessary optimality conditions we transform the time-optimal control problem (18) into an equivalent control problem on the unit time interval [0, 1] in which the final time T occurs as a control parameter beside the control function $u(.)$ and the control parameters $d^0, d^1, \ldots, d^n$. We consider the unknown initial state of H_0 as a control, but the unknown final state of H_1 as an additional state condition for $x(.)$.

Let $(x_0, u_0, d_0^0, d_0^1, \dots, d_0^n, T_0)$ be optimal. Then there exists a solution y (in the sense of evolution equation) of the boundary value problem

$$-\frac{\partial y}{\partial t}(t,s) - \sum_{i=1}^{n} \frac{\partial}{\partial s_i} \frac{\partial}{\partial x} a_i\left(t, s, \frac{\partial x_0}{\partial s_i}(t,s), d_0^i\right) \frac{\partial y}{\partial s_i}(t,s)$$

$$+ \frac{\partial}{\partial x} a_0(t, s, x_0(t,s), d_0^0)\, u_0(t,s)\, y(t,s) = 0\,, \qquad t \in [0, T_0],\ s \in \Omega\,,$$

$$y(t,s) = 0\,, \qquad t \in [0, T_0],\ s \in \partial\Omega\,,$$

such that

$$a_0(t, s, x_0(t,s))\, y(t,s)\, u_0(t,s) = \max_{u_1(s) \leqq u \leqq u_2(s)} a_0(\dots)\, y(t,s)\, u \qquad \text{a.e. on } [0, T] \times \Omega\,,$$

$$\int_0^{T_0} \int_\Omega \frac{\partial}{\partial d} a_i\left(t, s, \frac{\partial x_0}{\partial s_i}(t,s), d_0^i\right) \frac{\partial y}{\partial s_i}(t,s)\, \mathrm{d}s\, \mathrm{d}t\, d_0^i$$

$$= \max_{d \in [d_1^i, d_2^i]} \int_0^{T_0} \int_\Omega \frac{\partial}{\partial d} a_i\left(t, s, \frac{\partial x_0}{\partial s_i}(t,s)\, d_0^i\right) \frac{\partial y}{\partial s_i}(t,s)\, \mathrm{d}s\, \mathrm{d}t\, d\,, \qquad i = 1, \dots, n\,,$$

$$\int_0^{T_0} \int_\Omega \frac{\partial}{\partial d} a_0(t, s, x_0(t,s), d_0^0)\, u_0(t,s)\, y(t,s)\, \mathrm{d}s\, \mathrm{d}t\, d_0^0$$

$$= \max_{d \in [d_1^0, d_2^0]} \int_0^{T_0} \int_\Omega \frac{\partial}{\partial d} a_0(\dots)\, u_0(t,s)\, y(t,s)\, \mathrm{d}s\, \mathrm{d}t\, d\,,$$

$$\int_\Omega y(0,s)\, x_0(0,s)\, \mathrm{d}s = \min_{h \in H_0} \int_\Omega y(0,s)\, h(s)\, \mathrm{d}s\,,$$

$$\int_\Omega y(T_0,s)\, x_0(T_0,s)\, \mathrm{d}s = \max_{h \in H_1} \int_\Omega y(T_0,s)\, h(s)\, \mathrm{d}s\,,$$

$$\int_0^{T_0} \int_\Omega \left\{ \sum_{i=1}^{n} \left(\frac{\partial}{\partial t} a_i\left(t, s, \frac{\partial x_0}{\partial s_i}(t,s), d_0^i\right) t + a_i(\dots) \right) \frac{\partial y}{\partial s_i}(t,s) \right.$$

$$\left. + \left(\frac{\partial}{\partial t} a_0(t, s, x_0(t,s), d_0^0)\, t + a_0(\dots) \right) u_0(t,s)\, y(t,s) \right\} \mathrm{d}s\, \mathrm{d}t \geqq 0\,.$$

For more details see [13].

4. References

1 Bittner, L.: Necessary optimality conditions for a model of optimal control processes. Banach Center Publ. **1** (1975), 25–32.

2 –: On optimal control of processes governed by abstract functional, integral and hyperbolic differential equations. Math. Operationsforsch. Statist. **6** (1975), 107–134.

3 –: Ein Modell für eine Klasse von Aufgaben optimaler Steuerung. ZAMM **58** (1978), 251–260.

4 Gajewski, H.; Gröger, K.; Zacharias, K.: Nichtlineare Operatorgleichungen und Operatordifferentialgleichungen. Akademie-Verlag, Berlin 1974.

5 Kampowsky, W.: Optimalitätsbedingungen für Prozesse in Evolutionsgleichungen 1. Ordnung. ZAMM **61** (1981), 501–512.

6 –: Optimalitätsbedingungen für Prozesse in Evolutionsgleichungen n-ter Ordnung. Z. Anal. u. Anwend. **1** (1982), 77–94.

7 –: Zur optimalen Steuerung in Dirichletschen Randbedingungen bei Evolutionsgleichungen. ZAMM **66** (1986), 395–402.

8 –: Zur optimalen Steuerung in Neumannschen Randbedingungen bei Evolutionsgleichungen. ZAMM **66** (1986), 477–481.

9 –: On optimal boundary control in evolution equations. Materialy IX Konferencji Szkoleniowej z Teorii Zagadnien Estremalnych. Uniwersytaet Lodzki, Instytut Matematyki, Lodz 1988, pp. 102–112.

10 –: On existence of optimal solutions for control problems in evolution equations. Materialy X Konferencji Szkoleniowej z Teorii Zagadnien Ekstremalnych, Uniwersytaet Lodzki, Instytut Matematyki, Lodz 1989, pp. 158–171.

11 –: Zur Approximation von Steueraufgaben mit Evolutionsgleichungen. Wiss. Z. TH Leuna-Merseburg **31** (1989), 750–756.

12 –: Optimalitätsbedingungen und Existenzaussagen für Steueraufgaben mit Evolutionsgleichungen und Rand-Anfangswertproblemen für parabolische Differentialgleichungen im Rahmen der Theorie monotoner Operatoren. Diss. B, Ernst-Moritz-Arndt-Universität Greifswald 1988.

13 –: Time-optimal control in nonlinear evolution equations. (Submitted to Proceedings of the 14th IFIP Conference on System Modelling and Optimization, Leipzig, GDR, July, 3–7, 1989).

14 Lions, J. L.: Quelques méthodes de résolution des problèmes aux limites non linéaires. Dunod, Gauthier-Villars, Paris 1969.

Received January 11, 1990, revised April 9, 1990

Address: Dr. sc. Winfried Kampowsky, Ernst-Moritz-Arndt-Universität Greifswald, Fachbereich Mathematik, Friedrich-Ludwig-Jahn-Str. 15a, O-2200 Greifswald, Germany

SHORT COMMUNICATIONS

ZAMM · Z. angew. Math. Mech. 71 (1991) 7/8, 289–291 Akademie Verlag

KLEPP, H. J.

Approximate Solutions, Numerical Errors and Chaos of the Logistic Equation

Dedicated to the memory of LOTHAR COLLATZ

MSC (1980): 65Q05

1. Introduction

The standard logistic differential equation is a model, which describes the growth in presence of invariable restraining factors (crowding, limited resources) [1]. The general solution of this first order differential equation can be determined by quadratures.

The simplest example of a dynamic model that exhibits chaotic dynamic is the logistic difference equation. Transition from stable points to a bifurcating hierarchy of stable cycles and to chaotic fluctuations is described in nearly all references, which deal with chaotic dynamics (for example [2], [3], [4]).

The logistic difference equation can be considered as an equation for the determination of approximate solutions of the logistic differential equation, when Euler's polygon method is used. As mentioned in [5] and [6], this method is a crude approximate method which only should be used when no great accuracy is required, when only small integration steps are used, and when only a small number of integration steps are involved.

Because the exact solution of the logistic differential equation is known, it is possible to compare this solution with the solutions obtained by Euler's polygon method, i.e. by the logistic difference equation, when different integration steps are used. This fact allows to correlate phenomena as bifurcation of solutions, period multiplying and chaotic behaviour with the values of the integration steps and with numerical errors of the approximate solutions.

2. The logistic differential equation

The logistic differential equation $dy/dt = ay(1 - y/c)$ [1], with $a > 0$, $c > 0$ and $y > 0$, will be considered in the form

$$du/d\tau = f(u), \qquad f(u) = u(1 - u), \tag{1}$$

where $u = y/c > 0$ and $\tau = at$. The singular points of this equation, i.e. $f(u^*) = 0$, are $u^* = 0$ and $u^* = 1$. The stability of these singular points can be determined with $u = u^* + \xi$, $|\xi| \ll 1$, $f(u) \approx f(u^*) + \xi f'(u^*)$, $f'(u^*) = 1 - 2u^*$, from $d\xi/d\tau = \xi(1 - 2u^*)$. For $u^* = 0$ results $d\xi/d\tau = \xi$, i.e. the singular point $u^* = 0$ is unstable. For $u^* = 1$ results $d\xi/d\tau = -\xi$, i.e. the singular point $u^* = 1$ is stable.

The exact solution of equation (1) for initial conditions $\tau = \tau_i$, $u = u(\tau_i)$ is

$$u = \frac{1}{1 + [1/u(\tau_i) - 1] \cdot \exp[-(\tau - \tau_i)]} = u(\tau; \tau_i, u(\tau_i)). \tag{2}$$

For initial conditions $u(\tau_i) < 1$ or $u(\tau_i) > 1$ the solutions increase or decrease monotonously to the stable value $u = 1$.

3. The logistic difference equation

Approximate solutions of differential equation (1) can be determined by Euler's polygon method [5], [6]. Denoting $\tau_k = \tau_0 + k\,\Delta\tau$, $f_k = f(u_k) = u_k(1 - u_k)$, $k = 0, 1, 2, \ldots$, the formula for the calculation of the next u_{k+1} from u_k is

$$u_{k+1} = u_k + f_k \Delta\tau, \tag{3}$$

which leads to

$$u_{k+1} = (1 + \Delta\tau)\, u_k[1 - u_k\, \Delta\tau/(1 + \Delta\tau)]. \tag{4}$$

The *nondimensional logistic difference equation*, usually considered as an example for chaotic dynamics [3], is

$$x_{k+1} = \lambda x_k(1 - x_k). \tag{5}$$

A comparison between (4) and (5) shows that with

$$\lambda = 1 + \Delta\tau \tag{6}$$

and $x_k = u_k(\lambda - 1)/\lambda$ these equations are identical. Thus, both equations will behave similarly. For $\Delta\tau > 0$ results $\lambda > 1$.

In what follows, the logistic difference equation will be considered in the form

$$u_{k+1} = g(u_k), \qquad g(u_k) = \lambda u_k[1 - u_k(\lambda - 1)/\lambda]. \tag{7}$$

Because $\lambda = 1 + \Delta\tau$, the question arises, whether solutions of equation (7) for large λ-values, i.e. for large values of the integration step $\Delta\tau$, can be considered as approximate solutions of differential equation (1).

In [5] the error in the approximate value u_k is defined as the quantity

$$\varepsilon_k = u_k - u(\tau_k). \tag{8}$$

For the continuous function $f(u)$, bounded in a domain of the real plane (τ, u), which satisfies a Lipschitz condition there, and which possesses bounded partial derivatives, the "truncation" error estimation, carried out for the polygon method, shows that at a point τ_k the error limits tend to zero as $\Delta\tau \to 0$. When the domain $u \in (0, 2)$, $\tau \geqq 0$, is considered, the error estimate established in [5] leads to $|\varepsilon_k| \leqq \Delta\tau\{\exp[3(\tau_k - \tau_0)] - 1\}$.

Thus, large λ-values, i.e. large values of the integration steps $\Delta\tau$, will contradict this requirement, and for reasons of "local" errors, the solutions of equation (7) can hardly be considered as approximate solutions of equation (1).

The fixed points of equation (7), i.e. $u_k^* = g(u_k^*)$, are $u_k^* = 0$ and $u_k^* = 1$. They coincide with the singular points of equation (1).

Because the exact solutions of equation (1) tend to the stable singular point $u^* = 1$, the convergence properties of the approximate solutions, i.e. of the successive values u_{k+i}, $i = 1, 2, \ldots$, of equation (7), are also of interest. The "global" behaviour of these solutions, i.e. their convergence and the values to which they converge, can also be a criterion to decide whether results of equation (7) can still be considered as acceptable approximate solutions of equation (1), or not.

Let

$$|u_{k+2} - u_{k+1}| < |u_{k+1} - u_k| \tag{9}$$

be a criterion for convergence of the solutions of equation [7]. This condition leads to

$$h(u_k, \lambda) < 1, \qquad h(u_k, \lambda) = |[1 - (\lambda - 1)\, u_k]\, [\lambda - (\lambda - 1)\, u_k]|. \tag{10}$$

Figures 1 and 2 present $h = h(u_k, \lambda)$ for $\lambda < 3$ and for $\lambda > 3$. The points v_i, $i = 0, 1, \ldots, 6$, have the u_k-coordinates

$$v_0 = \frac{\lambda + 1}{2(\lambda - 1)}, \tag{11}$$

$$\left.\begin{matrix} v_1 \\ v_2 \end{matrix}\right\} = \frac{1}{2(\lambda - 1)}\,[\lambda + 1 \mp \sqrt{(\lambda - 1)^2 + 4}], \tag{12}$$

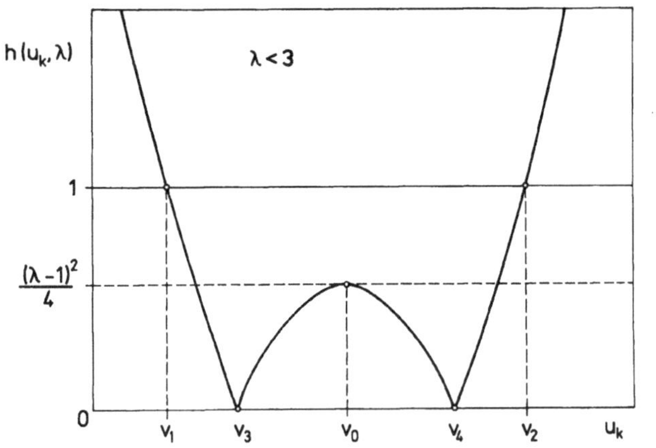

Fig. 1. Plot of the function $h = h(u_k, \lambda)$ for $\lambda < 3$.

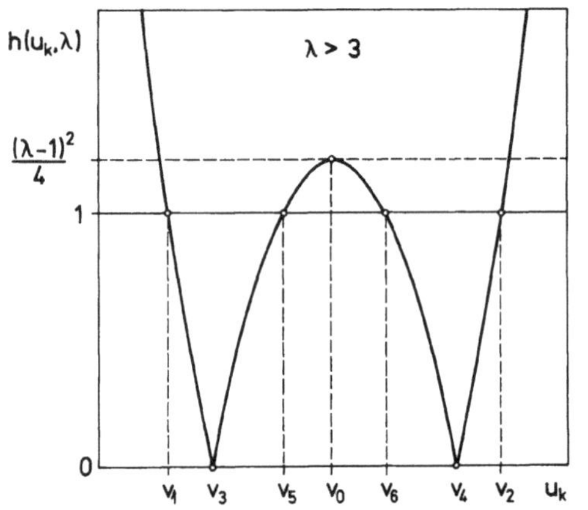

Fig. 2. Plot of the function $h = h(u_k, \lambda)$ for $\lambda > 3$.

$$v_3 = \frac{1}{\lambda - 1}, \qquad v_4 = \frac{\lambda}{\lambda - 1}, \tag{13}$$

$$\left.\begin{matrix} v_5 \\ v_6 \end{matrix}\right\} = \frac{1}{2(\lambda - 1)} \left[\lambda + 1 \mp \sqrt{(\lambda + 1)(\lambda - 3)}\right]. \tag{14}$$

The values $u_k = v_i = v_i(\lambda)$, $i = 3, 4, 5, 6$, have particular properties. There holds: $g(v_3) = 1$, $g(v_4) = 0$, $g(v_5) = v_6$, $g(v_6) = v_5$. Figure 3 presents the curves $u_k = v_i = v_i(\lambda)$, $i = 1, 3, 4, 5, 6$.

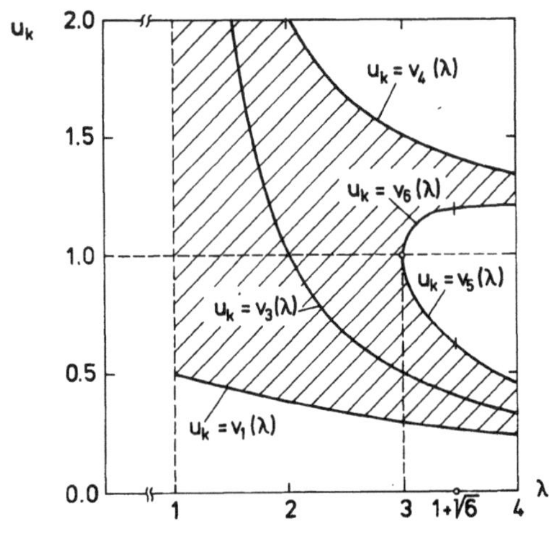

Fig. 3. Plots of the functions $u_k = v_i(\lambda)$, $i = 1, 3, 4, 5, 6$.

The value $u_k = u_k^* = 0$ is situated outside the domain of u_k-values with property (9). Thus, the fixed point $u_k^* = 0$ is unstable for all λ-values, similar to the behaviour of solution (2).

In the domain $\lambda \in (1, 3)$ the value $u_k = u_k^* = 1$ is situated inside the domain of u_k-values, which satisfy condition (9). Thus, in this domain the u_{k+i}-values, $i = 1, 2, \ldots$, tend to the fixed point $u_k^* = 1$, similar to the behaviour of solution (2).

In the domain $\lambda > 3$ the fixed point $u_k^* = 1$ belongs no longer to the domain of u_k-values with property (9), i.e. the u_{k+i}-values, $i = 1, 2, \ldots$, will not converge to $u_k^* = 1$. Thus, the solutions of equation (7) can no longer be considered as acceptable approximate solutions of equation (1). For $\tau = \tau_k$, which correspond to $u(\tau_k) \approx 1$, the quantities $\varepsilon_k = u_k - 1$ are errors of the approximate solutions. Thus, periodical or irregular behaviour of the solutions of equation (7) are properties of these numerical errors.

The values $v_5 = v_5(\lambda)$ and $v_6 = v_6(\lambda)$, which satisfy $g(v_5) = v_6$ and $g(v_6) = v_5$, are the fixed points, when doubling of solutions in equation (7) occurs. These values belong to the solutions of equation $u_k = g(g(u_k))$ [5]. For the analysis of their stability a perturbed starting value $\bar{v}_5 = v_5 + \zeta$, $|\zeta| \ll 1$ will be considered. From $g(\bar{v}_5) \approx g(v_5) + \zeta g'(v_5) = v_6 + \zeta_1 = \bar{v}_6$, and $g(\bar{v}_6) \approx g(v_6) + \zeta_1 g'(v_6) = v_5 + \zeta_2$, and condition $|\zeta_2| < |\zeta|$, results: for $\lambda \in (3, 1 + \sqrt{6})$ the fixed points v_5 and v_6 are stable.

4. Numerical examples

Figure 4 presents from initial value $\tau_i = 0$, with $u(0) = 0.8$, till $\tau = 16.2$ the exact solution given by equation (2) (curve 1), and polygons 2 and 3 obtained with equation (7). Polygon 2 results for $\Delta\tau = 0.9$ ($\lambda = 1.9$), and can be considered as a very crude approximation of curve 1. Polygon 3 results for $\Delta\tau = 1.8$ ($\lambda = 2.8$). For this large integration step the pattern of this solution differs significantly from the exact solution. These curves only have the property in common that they tend to the value $u = 1$.

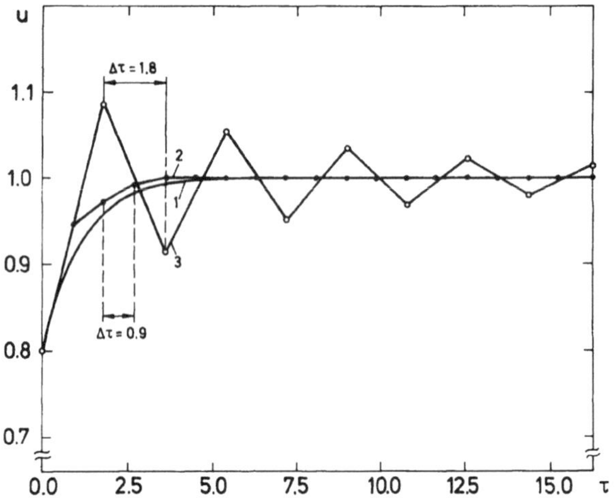

Fig. 4. Exact solution (curve 1) and approximate solutions for integration steps $\Delta\tau = 0.9$ (polygon 2) and $\Delta\tau = 1.8$ (polygon 3) of the logistic differential equation.

Figures 5 and 6 present for initial conditions $\tau_i = 0$, $u(0) = 0.8$, the exact solution given by equation (2) (curve 1) and polygons 4 and 5 obtained with equation (7). The results calculated with equation (7) are completed with the curves $u = u(\tau; \tau_k, u_k)$ given by equation (2) for $\tau \in [\tau_k, \tau_{k+1}]$.

Polygon 4 in Figure 5 results from equation (7) for $\Delta\tau = 2.2$ ($\lambda = 3.2$). After several integration steps the u_{k+i} -values , $i = 1, 2, \ldots$, alternate between $v_5(3.2) = 0.7462$ and $v_6(3.2) = 1.1628$. Because the exact solution in this domain of τ-values is $u \approx 1$, this bifurcated solution can be interpreted as an approximate solution with constant numerical errors alternating between -0.2538 and 0.1628.

Polygon 5 in Figure 6 results from equation (7) for $\Delta\tau = 2.7$ ($\lambda = 3.7$). For this λ-value the u_{k+i}-values alternate irregularly. This kind of behaviour is named *chaotic* [2]. The

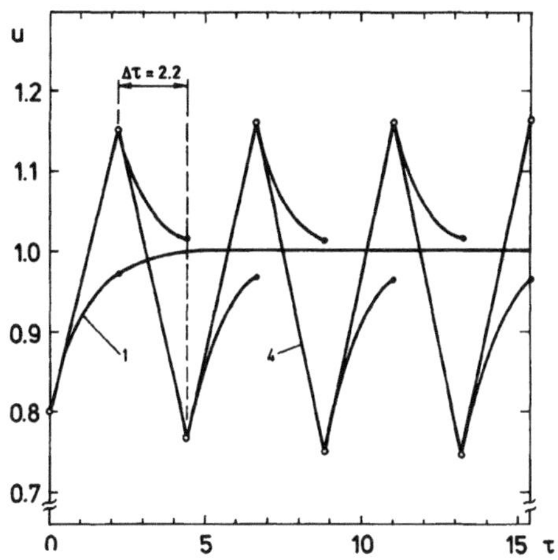

Fig. 5. Exact solution (curve 1), approximate solution for integration step $\Delta\tau = 2.2$ (polygon 4) and curves, which present the exact solutions with initial conditions τ_k, u_k for $\tau \in [\tau_k, \tau_{k+1}]$, of the logistic differential equation.

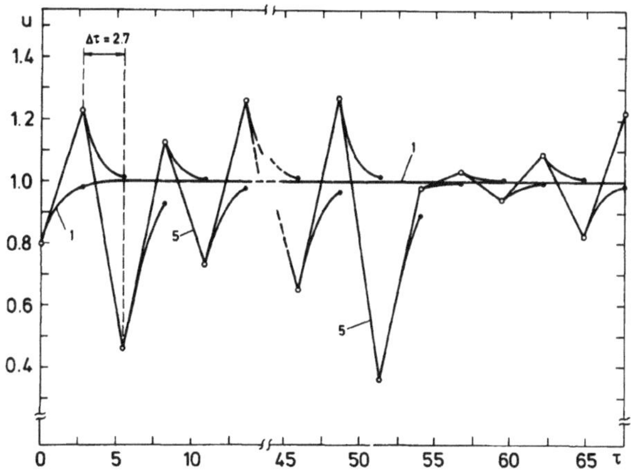

Fig. 6. Exact solution (curve 1), approximate solution for integration step $\Delta\tau = 2.7$ (polygon 5) and curves, which present the exact solutions with initial conditions τ_k, u_k for $\tau \in [\tau_k, \tau_{k+1}]$, of the logistic differential equation.

corresponding u_{k+i}-values , $i = 1, 2, \ldots$, can be interpreted as approximate solutions with numerical errors, which alternate irregularly.

Bifurcation and chaotic behaviour of solutions known for equation (7), which arise for integration steps $\Delta\tau > 2$, appear as properties of numerical errors of approximate solutions, provided the results of equation (7), determined with such large integration steps and after a large number of integration steps, can be still considered as approximate solutions of differential equation (1).

5. Conclusions

When the logistic difference equation (7) is considered as an equation for the calculation of approximate solutions of the logistic differential equation (1) by Euler's polygon method, then bifurcation of solutions, and transition to periodic or chaotic solutions, observed on equation (7), appear as properties of numerical errors in case of inappropriate (i.e. too large) integration steps. Thus, it is questionable, whether or not properties of numerical errors can be interpreted as real phenomena of the dynamical system governed by the considered differential equation.

References

1 LUENBERGER, D. G.: Introduction to dynamic systems. Theory, models, and applications. John Wiley & Sons, New York 1979, pp. 317, 318.

2 MAY, R. M.: Simple mathematical models with very complicated dynamics. Nature **261** (1976), 459–467.

3 MOON, F. M.: Chaotic vibrations. An introduction for applied scientists and engineers. John Wiley & Sons, New York 1987, pp. 25–27, 169.

4 SEIFRITZ, W.: Wachstum, Rückkopplung und Chaos. Eine Einführung in die Welt der Nichtlinearitäten und des Chaos. Carl Hanser Verlag, München 1989, pp. 19–21, 41–65

5 COLLATZ, L.: The numerical treatment of differential equations. Springer Verlag, Berlin 1966, p. 53, 57–59.

6 KORN, A. G.; KORN, T. M.: Mathematical handbook for scientists and engineers. McGraw-Hill Book Company, Inc., New York 1961.

Received May 31, 1990

Address: Dr. rer. nat. habil. HORST J. KLEPP, Ruhr-Universität Bochum, Institut für Mechanik, Universitätsstraße 150, W-4630 Bochum 1, Germany

ZAMM · Z. angew. Math. Mech. **71** (1991) 7/8, 291–292 Akademie Verlag

WETTERLING, W.

Application of Chebyshev Approximation to Curve Fitting

Dedicated to the memory of LOTHAR COLLATZ

MsC (1980): 65D07, 65D10

1. Introduction

The standard methods of curve and surface fitting are based upon least squares approximation of given data. An alternative is the principle of uniform (Chebyshev) approximation. This approach can be advantageous and even essential in certain technical applications, such as the engineering problem described in section 2. The general problem of uniform curve fitting to given data with some generalizations and two special cases are discussed in section 3.

It should be noted that these problems are different from those of standard Chebyshev approximation where a given function is approximated by an element of a linear or nonlinear class of functions. However, as will be shown in section 4, certain principles known from approximation theory such as optimality conditions and alternation properties can be retrieved here.

2. An approximation problem in railway engineering

The work described in this section has been done in collaboration between the Department for Infrastructure of the Dutch Railway Company (Nederlandse Spoorwegen N.V.) and the Systems and Optimization Group of the Applied Mathematics Department at the University of Twente. Here we give a short account of the technical and mathematical aspects of the problem. More details can be found in the graduate thesis of E. JONKER [2].

The technical problem is as follows: It is required that the upper contour of the rails on a railway bridge is a C^2-curve (second derivatives bounded and continuous). The permitted maximum speed on the bridge depends upon a lower bound for the radius of curvature of that curve. In order to achieve the required smoothness, after construction of a new railway bridge or in the case of renewal of tracks on existing bridges the irregularities in the bridge surface must be compensated, e.g. by adjusting the heights of the rail supports.

In the mathematical model the upper contour of the rails is represented by a cubic spline function which approximates a set of data (x_i, y_{Li}, y_{Ri}), $i = 0, 1, \ldots, m$, describing the bridge surface. The x_i are the horizontal coordinates (locations of rail supports), $A = x_0 < x_1 < \ldots < x_m = B$, and y_{Li}, y_{Ri} are vertical coordinates for the left and right rail (reference heights, obtained by measurements of the bridge surface). A set of (fixed) knots p_j is chosen, $A = p_0 < p_1 < \ldots < p_k = B$ (in practice k between $m/20$ and $m/8$, each p_j coincident with an x_i).

The approximating cubic spline function with knots p_j is denoted by $s = s(x, a_1, \ldots, a_{k+3})$. The linear parameters a_j have to be determined such that the following conditions are satisfied: The deviation between $s(x_i, a_j)$ and the reference heights y_{Li} and y_{Ri}

should be less than or equal to a tolerance h,

$$\left.\begin{array}{l}|s(x_i, a_j) - y_{Li}| \leqq h \\ |s(x_i, a_j) - y_{Ri}| \leqq h\end{array}\right\}, \quad i = 1, 2, \ldots, m. \tag{1}$$

The radius of curvature should be bounded from below by some r_0,

$$r = [1 + (\partial s/\partial x)^2]^{1/2}/|\partial^2 s/\partial x^2| \geqq r_0, \quad A \leqq x \leqq B. \tag{2}$$

(we allow a zero denominator). Moreover, boundary conditions for

$$s, \quad \partial s/\partial x \quad \text{and} \quad \partial^2 s/\partial x^2, \quad x = A, B, \tag{3}$$

possibly also values in some intermediate points, are prescribed.

The nonlinear condition (2) in all points of the interval [A, B] can be reduced to a linear condition on the finite set of knots. (2) will be satisfied if

$$1/|\partial^2 s/\partial x^2| \geqq r_0, \quad A \leqq x \leqq B, \tag{2a}$$

holds, and sufficient for (2a) is

$$|\partial^2 s/\partial x^2| \leqq 1/r_0, \quad x = p_0, p_1, \ldots, p_k, \tag{2b}$$

since the second derivative of a cubic spline is a linear spline and therefore has its local extrema in the knots.

The original problem is now reduced to solving a finite system of linear inequalities (1), (2b) and equations (3). In the actual computations as described in [2] the problem is solved as an approximation problem by taking $h = c_1\varepsilon$, $l/r_0 = c_2\varepsilon$ with constants c_1, c_2 and minimizing ε under the conditions (1) and (2b) after a suitable matrix transformation to ensure (3). A modified simplex method, essentially STIEFEL's algorithm [3], is used to solve this problem of discrete linear Chebyshev approximation. The numbers c_1 and c_2 are adapted appropriately in order to find a solution satisfying all requirements, possibly with an r_0 too small to allow maximum speed. In such cases E. JONKER's program gives information how to obtain a better surface by suitable modification of the data (e.g. grinding off some sections of the surface, or using leveling pads in the rail supports). The program can also calculate spline approximations for railway bridges with prescribed banking in horizontal curves.

The program calculated successfully several practical examples: A bridge over the Strobosservaart ($m = 46$, $r_0 = 8.8$ km), a steel bridge near Veendam ($m = 42$, $r_0 = 4.6$ km) and a concrete fly-over near Lage Zwaluwe ($m = 589$, $r_0 = 12.3$ km).

3. Uniform approximation by curves in $\mathbb{R}^n$

We consider a generalization of the problem above. Suppose a finite set of points $\{y_1, y_2, \ldots, y_m\} \subset \mathbb{R}^n$ and a family of curves $x(t, a)$ in $\mathbb{R}^n$, $t \in \mathbb{R}$, $a \in A \subset \mathbb{R}^k$, are given. The problem is to find an $a^* \in A$ such that the curve $x(t, a^*)$ is a best approximation to the points y_i. That means we have to determine, with $\|.\|$ the Euclidean norm,

$$\min_a \max_i \min_t \|y_i - x(t, a)\|. \tag{4}$$

Constraints like the condition for the radius of curvature above can be given, possibly included in the definition of the set A.

As an applied example we mention the problem where both the vertical and the horizontal contour of a railway section are modeled as smooth curves. There are several other possible technical applications, for instance in robotics.

We can regard problem (4) as a problem of curve fitting in the Chebyshev sense, where the maximum of the distances and not as usually the sum of squares is minimized. Even more general problems might be worth considering, for instance the semi-infinite problem with a general compact set Y instead of the finite set $\{y_1, \ldots, y_m\}$, or the problem where t is in a set $T \subset \mathbb{R}^d$ ($d = 1$ curve fitting, $d = 2$ surface fitting).

We mention two special cases of this general problem. In the first case $x(t, a) = a =$ const. is independent of t. For a given finite set of points $\{y_1, \ldots, y_m\}$ the unique point a^* is determined which minimizes $\max_i \|y_i - a\|$. This problem is known as the *minimum covering sphere problem*. The optimal point a^* is called the *Chebyshev center* of the set $\{y_1, \ldots, y_n\}$. In [1] the problem is written as a convex optimization problem and an algorithm based on WOLFE's duality theory is given which determines the Chebyshev center in a finite number of steps.

The second case which will be the subject of a forthcoming paper is approximation by a straight line, $x(t) = c + bt$, $c, b \in \mathbb{R}^n$, $t \in \mathbb{R}$. Here the approximation problem can be written as a non-convex optimization problem. Several local minima are possible. An iterative method has been developed which determines alternately an improved c using the finite algorithm for the Chebyshev center and an improved b as solution of a quadratic optimization problem.

4. Optimality conditions

Let $x(t, a)$ in (4) be sufficiently smooth. If $t = t_i$ minimizes $\|y_i - x(t, a)\|$ then

$$\left(\frac{dx}{dt}(t_i, a)\right)^{\mathrm{T}} (y_i - x(t_i, a)) = 0. \tag{5}$$

We assume that the conditions of the implicit function theorem are satisfied and hence (5) has a solution $t_i = t_i(a)$ which is continuously differentiable in some neighborhood of a^*. We can write (4) as an optimization problem:
Minimize $F(r, a) = r$, subject to

$$\left.\begin{array}{l}g_i(r, a) = r - (y_i - x(t_i(a), a))^{\mathrm{T}} (y_i - x(t_i(a), a)) \geqq 0, \\ i = 1, \ldots, m.\end{array}\right\} \tag{6}$$

The 1st order optimality conditions for this problem are

$$\left.\begin{array}{l}1 - \sum\limits_{i \in J} \lambda_i = 0, \quad \lambda_i \geqq 0, \quad i \in J, \\ \sum\limits_{i \in J} \lambda_i \left(\dfrac{\partial x(t_i, a)}{\partial a}\right)^{\mathrm{T}} (y_i - x(t_i, a)) = 0,\end{array}\right\} \tag{7}$$

where the set J contains the indices of active constraints in (6) and λ_i are the Lagrange multipliers. Note that due to (5) a term containing dx/dt could be omitted. Geometrically (7) describes how the active points with distance $\sqrt{r}$ are situated along the curve $x(t, a)$.

As an example consider approximation by a straight line:

$$x(t) = c + bt, \quad a = \begin{pmatrix} c \\ b \end{pmatrix} \in \mathbb{R}^{2n}, \quad \frac{\partial x(t, a)}{\partial a} = (I, tI). \tag{8}$$

With $t_i = b^{\mathrm{T}}(y_i - c)/b^{\mathrm{T}}b$ condition (7) becomes

$$\left.\begin{array}{l}\sum\limits_{i \in J} \lambda_i = 1, \quad \lambda_i \geqq 0, \quad i \in J, \\ \sum\limits_{i \in J} \lambda_i(y_i - c - bt_i) = 0, \quad \sum\limits_{i \in J} \lambda_i t_i(y_i - c - bt_i) = 0.\end{array}\right\} \tag{9}$$

It is easy to see that in the case $n = 2$ (approximation by a straight line in the plane) this condition implies the alternation property for Chebyshev approximation by first degree polynomials. Evidently we can consider (7) as a generalized alternation property.

Acknowledgements: The author wishes to thank ir. E. JONKER for his assistance during the work described in section 2 and ing. T. MIEDEMA, Nederlandse Spoorwegen N.V., for the guidance of E. JONKER's work and for helpful comments during the preparation of this paper.

References

1 ELZINGA, D. J.; HEARN, D. W.: The minimum covering sphere problem. Management Sci. 19 (1972), 96–104.
2 JONKER, E.: Geautomatiseerde bepaling van een voldoende gladde spline voor het spoorstavenbeloop in het verticale vlak. Doctoraalverslag, Department of Applied Mathematics, University of Twente 1987.
3 STIEFEL, E.: Note on Jordan elimination, linear programming and Tschebyscheff-approximation. Numerische Math. 2 (1960), 1–17.

Received May 28, 1990

Address: Prof. Dr. WOLFGANG WETTERLING, Twente University of Technology, Department of Applied Mathematics, Postbus 217, NL-7500 AE Enschede, The Netherlands

ZAMM · Z. angew. Math. Mech. 71 (1991) 7/8, 293–295 Akademie Verlag

Kwapisz, M.

On the Error Estimate of Approximate Solutions of the Discrete Volterra Equations of k-th Order

Dedicated to the memory of Lothar Collatz

MSC (1980): 41A65, 65M15, 65N15

1. The problems of the theory of recurrence equations, the theory of iterative methods of solving fixed point equations as well as the problems related to the convergence of numerical methods of solving initial value problems for ODE's or PDE's lead to discussions on questions concerning the error of approximate solutions of discrete equations.

The aim of the present paper is to give an extension of the abstract approach presented in [6] (consult [5, 7] for the problems related with). Here we deal with equations of k-th order defined by isotone-antitone mappings instead of isotone ones considered in [6].

Let $(X, \leqq)$ be a linear partially ordered space. We accept the usual terminology concerning concepts connected with the relation "$\leqq$" (see for example [1–3, 8, 9]). Suppose that X is a vector lattice, i.e. for any $x, y \in X$ there exists $\sup (x, y) \in X$ (see [8, 9]).

Define the absolute value of the element $x \in X$, $|x| = \sup (x, -x)$. It is known that the absolute value $|\ldots|$ has the properties:

$$|x| \geqq 0, \quad |x| = 0 \quad \text{iff} \quad x = 0,$$
$$|x + y| \leqq |x| + |y|,$$
$$|\alpha x| = |\alpha|\,|x|, \qquad \alpha \text{ a scalar},$$
$$|x| \leqq y \quad \text{iff} \quad -y \leqq x \leqq y,$$

(for proof see [1, 8, 9]). Let $N = \{0, 1, \ldots\}$. For fixed $-p, q \in N$ put $Z(p, q) = \{p, p + 1, \ldots, q\}$; the case when $-p, q = +\infty$ is also accepted.

Put $X^+ = \{x \mid x \in X, x \geqq 0\}$. Let $\Phi = \Phi(p, q)$ denote the set of all mappings from $Z = Z(p, q)$ to X; Φ^+ denotes the same with X replaced by X^+. The natural partial order in Φ is induced by that one in X.

For given positive integer k consider a mapping F: $Z(0, q - k) \times \Phi \times \Phi \to X$. We say that F has the *Volterra property* (see [5]) when $x, x', y, y' \in \Phi$ and $x(i) = x'(i)$, $y(i) = y'(i)$ for $i \in Z(p, n + k - 1)$ imply

$$F(n, x, y) = F(n, x', y'), \qquad n \in Z(0, p - k).$$

For such F we write $F(n, x, y)$ also as $F(n, [x] (n + k - 1), [y] (n + k - 1))$.

We say that the mapping F is *isotone-antitone* if for any $x, x', y, y' \in \Phi$ the conditions $x \leqq x'$ and $y' \leqq y$ imply the inequality

$$F(n, x, y) \leqq F(n, x', y'), \qquad n \in Z(0, q - k).$$

Now for given $a \in \Phi(p, k - 1)$ consider the following discrete initial-value problem:

$$x(n + k) = F(n, [x] (n + k - 1), [x] (n + k - 1)),$$
$$n \in Z(0, q - k), \tag{1}$$
$$x(i) = a(i), \qquad i \in Z(p, k - 1). \tag{2}$$

Equation (1) is called the *Volterra discrete equation of k-th order*. Because of the Volterra property of F, there always exists exactly one solution $x \in \Phi$ of the problem (1)–(2).

Let $y \in \Phi$ be an (s, r)-approximate solution of the problem (1)–(2), i.e.,

$$|y(n + k) - F(n, [y] (n + k - 1), [y] (n + k - 1))| \leqq r(n),$$
$$n \in z(0, q - k), \tag{3}$$
$$|y(i) - a(i)| \leqq s(i), \qquad i \in Z(p, k - 1), \tag{4}$$

for some $s \in \Phi(p, k - 1)$ and $r \in \Phi(0, q - k)$.

Our aim is to give an evaluation of the error $x - y$ between the exact solution x and the approximate solution y of the problem (1)–(2).

2. To get the mentioned error evaluation we use the following result on recurrent inequalities which is proved by a simple induction.

Theorem 1: *If the isotone-antitone mapping F has the Volterra property, x is defined by (1)–(2) and $u, v \in \Phi$ are sucht that*

$$u(n + k) \leqq F(n, [u] (n + k - 1), [v] (n + k - 1)),$$
$$n \in Z(0, q - k), \tag{5}$$
$$v(n + k) \geqq F(n, [v] (n + k - 1), [u] (n + k - 1)),$$
$$n \in Z(0, q - k), \tag{6}$$

and

$$u(i) \leqq a(i) \leqq v(i), \qquad i \in Z(p, k - 1), \tag{7}$$

then

$$u(i) \leqq x(i) \leqq v(i), \qquad i \in Z. \tag{8}$$

Proof: By (7) the relation (8) holds for $n \in Z(p, k - 1)$. Assume that it holds for all $i \in Z(p, n + k - 1)$, then

$$x(n + k) = F(n, [x] (n + k - 1), [x] (n + k - 1))$$
$$\leqq F(n, [v] (n + k - 1), [u] (n + k - 1)) \leqq v(n + k)$$

and

$$x(n + k) = F(n, [x] (n + k - 1), [x] (n + k - 1))$$
$$\geqq F(n, [u] (n + k - 1), [v] (n + k - 1)) \geqq u(n + k)$$

what means that (8) holds for all $i \in Z(p, n + k)$. This together with the induction rule proves that (8) holds for all $i \in Z$.

Now let us take $f, g \in \Phi$ defined by the conditions:

$$f(n + k) \geqq F(n, y + f, y - g) - F(n, y, y) + r(n),$$
$$n \in Z(0, q - k), \tag{9}$$
$$f(i) \geqq s(i), \qquad i \in Z(p, k - 1), \tag{10}$$

and

$$g(n + k) \geqq F(n, y, y) - F(n, y - g, y + f) + r(n),$$
$$n \in Z(0, q - k), \tag{11}$$
$$g(i) \geqq s(i), \qquad i \in Z(p, k - 1). \tag{12}$$

By (3) and (4) we have

$$F(n, y, y) - r(n) \leqq y(n + k) \leqq F(n, y, y) + r(n),$$
$$n \in Z(0, q - k), \tag{13}$$
$$y(i) - s(i) \leqq a(i) \leqq y(i) + s(i), \qquad i \in Z(p, k - 1). \tag{14}$$

Put $v = y + f$, $u = y - g$. Now by (13), (14) and the definition of f and g we get

$$v(n + k) = F(n, v, u) - F(n, v, u) + y(n + k) + f(n + k)$$
$$\geqq F(n, v, u) + [f(n + k) - F(n, v, u) + F(n, y, y) - r(n)].$$

We found that

$$v(n + k) \geqq F(n, v, u), \qquad n \in Z(0, q - k), \tag{15}$$

and

$$v(i) \geqq a(i), \qquad i \in Z(p, k - 1). \tag{16}$$

Similarly we obtain

$$u(n + k) \leqq F(n, u, v), \qquad n \in Z(0, q - k), \tag{17}$$
$$u(i) \leqq a(i), \qquad i \in Z(p, k - 1). \tag{18}$$

These together with the definition of x (given by (1), (2)) and Theorem 1 imply the following

Theorem 2: *If*

1) *the isotone-antitone mapping F has the Volterra property,*

2) *f and g are defined by the conditions* (9), (10) *and* (11), (12) *respectively,*

3) *for $x, y \in \Phi$ the conditions* (1), (2) *and* (3), (4) *hold, respectively, then*

$$-g(n) \leqq x(n) - y(n) \leqq f(n), \qquad n \in Z, \tag{19}$$

moreover,

$$|x(n) - y(n)| \leqq g^*(n), \qquad n \in Z, \tag{20}$$

when

$$g^*(n) = \sup\,[f(n), g(n)] \geqq 0, \qquad n \in Z.$$

Observe that f and g defined above depend in general on y which in many cases is not known; so they can not be evaluated effectively. To avoid this deficiency of our result we have to assume more about the mapping F. We have the following

Theorem 3: *If*

1) *the conditions* 1) *and* 3) *of Theorem 2 are satisfied,*

2) *there is a mapping $Q: Z(0, q-k) \times \Phi \times \Phi \to X$ having the Volterra property and such that*

$$F(n, t + z, t' - w) - F(n, t, t') \leqq Q(n, z, w) \tag{21}$$

for all $t, t' \in \Phi$ and $w, z \in \Phi^+$,

3) *the elements $f, g \in \Phi$ satisfy the relations*

$$f(n + k) \geqq Q(n, f, g) + r(n), \qquad n \in Z(0, q - k), \tag{22}$$

$$f(i) \geqq s(i), \qquad i \in Z(p, k - 1), \tag{23}$$

$$g(n + k) \geqq Q(n, g, f) + r(n), \qquad n \in Z(0, q - k), \tag{24}$$

$$g(i) \geqq s(i), \qquad i \in Z(p, k - 1), \tag{25}$$

then

$$-f(n) \leqq x(n) - y(n) \leqq g(n), \qquad n \in Z, \tag{26}$$

moreover

$$|x(n) - y(n)| \leqq g^*(n), \qquad n \in Z, \tag{27}$$

when

$$g^*(n) = \sup\,[f(n), g(n)] \geqq 0, \qquad n \in Z.$$

Proof: To prove the theorem it is enough to observe that f and g defined by the conditions (22)–(23) and (24)–(25) satisfy the requirements of Theorem 2 and so the assertion holds.

Observe that one can assume $f = h$ and $g = h$ if $h \in \Phi^+$ and h is defined by the conditions:

$$h(n + k) \geqq Q(n, h, h) + r(n), \qquad n \in Z(0, q - k), \tag{28}$$

$$h(i) \geqq s(i), \qquad i \in Z(p, k - 1). \tag{29}$$

3. In the general case when the mapping F has no isotone-antitone property one can get an error estimation if a stronger condition on F is assumed. In this case there is no need to consider the double dependence of F on the unknown variable x. We have

Theorem 4: *Assume that*

1) *the mapping $F: Z(0, q-k) \times \Phi \to X$ has the Volterra property,*

2) *there is a mapping $Q^*: Z(0, q-k) \times \Phi^+ \to X^+$ with the Volterra property, isotone with respect to the second variable and such that for every $z, w \in \Phi$,*

$$|F(n, z) - F(n, w)| \leqq Q^*(n, |z - w|), \qquad n \in Z(0, q - k), \tag{30}$$

3) *for x and y the conditions* (1)–(2) *and* (3)–(4) *hold with $F(n, x, x)$ and $F(n, y, y)$ replaced by $F(n, x)$ and $F(n, y)$ respectively,*

4) *the element $m \in \Phi^+$ is defined by the conditions*

$$m(n + k) \geqq Q^*(n, m) + r(n), \qquad n \in Z(0, q - k), \tag{31}$$

$$m(i) \geqq s(i), \qquad i \in Z(p, k - 1), \tag{32}$$

then

$$|x(n) - y(n)| \leqq m(n), \qquad n \in Z. \tag{33}$$

Proof: By the definition of x and y we have

$$x(n + k) = F(n, x)$$

and

$$F(n, y) - r(n) \leqq y(n + k) \leqq F(n, y) + r(n)$$

hence

$$F(n, y) - F(n, x) - r(n) \leqq y(n + k) - x(n + k) \leqq F(n, y) - F(n, x) + r(n).$$

Now by the use of (30) we get

$$|y(n + k) - x(n + k)| \leqq Q^*(n, |y - x|) + r(n), \quad n \in Z(0, q - k). \tag{34}$$

We also have

$$|y(i) - x(i)| \leqq s(i), \qquad i \in Z(p, k - 1). \tag{35}$$

Now Theorem 1 is applied to get the result of the theorem.

4. Let us notice that there is a plenty of special cases which can be considered. For some possible applications for the case $k = 1$ consult the paper [6]. Many stability results for recurrent equations appearing in the literature are special cases of the results mentioned above (clearly in this case it is assumed that $q = +\infty$). The same is true for error evaluations for approximations obtained by discretization processes for differential problems of evolutive type. For details with this respect consult the paper [4] and the references mentioned there. We observe that in this case the mappings F, x as well as r and s and also the numbers p and q depend on discretization parameters. It is not the aim of this short note to discuss details of the abstract theory developed here.

We conclude the paper with a simple illustration of the result of Theorem 3. Assume that $k = 1$, $X = R$ the set of real numbers, $p = 0$ and that (1)–(2) and (3)–(4) take the form

$$x(n + 1) = F(n, x(n), x(n)), \qquad n \in Z(0, q - 1),$$

$$x(0) = a,$$

$$|y(n + 1) - F(n, y(n), y(n))| \leqq r(n), \qquad n \in Z(0, q - 1),$$

$$|y(0) - a| \leqq s,$$

respectively. Assume now that (21) has the form

$$F(n, t + z, t' - w) - F(n, t, t') \leqq Lz + Pw,$$

for all $t, t' \in R$, $w, z \in R^+$ and some $L, P \in R^+$. Now f and g are sequences of real numbers defined by the inequalities

$$f(n + 1) \geqq Lf(n) + Pg(n) + r(n),$$

$$g(n + 1) \geqq Pf(n) + Lg(n) + r(n), \qquad n \in Z(0, q - 1),$$

$$f(0) \geqq s, \qquad g(0) \geqq s.$$

As it was mentioned both f and g can be replaced by h being the solution of the problem

$$h(n + 1) = (L + P)\,h(n) + r(n), \qquad h(0) = s.$$

One has

$$h(n) = (L + P)^n s + \sum_{i=0}^{n-1} (L + P)^{n-i-1} r(i)$$

or

$$h(n) = (L + P)^n s + r(1 - L - P)^{-1} (1 - (L + P)^n)$$

when $r(n) \equiv r$. Now the error evaluation (27) holds with g^* replaced by h defined above.

Observe that for the case when $X = R^m$ similar formulas can be derived. In this case L and P stand for the corresponding matrices.

References

1 BOHL, E.: Monotonie: Lösbarkeit und Numerik bei Operatorgleichungen. Springer, Berlin 1974.
2 COLLATZ, L.: Funktionalanalysis und Numerische Mathematik. Springer, Berlin 1964.
3 DEIMLING, K.: Nonlinear functional analysis. Springer, Berlin 1985.
4 KAMONT, Z.; KWAPISZ, M.: Difference methods for nonlinear parabolic differential-functional systems with initial boundary conditions of the Neumann type. Commentat. Math. 28 1989, 223–248.
5 KWAPISZ, M.: On discrete equations, inequalities and interval evaluations of fixed points of isotone-antitone mappings. In NICKEL, K. L. E. (ed.): Interval mathematics. Academie Press, New York 1980, pp. 397–406.
6 KWAPISZ, M.: On the error evaluation of approximate solutions of discrete equations. Nonlinear Analysis, TMA 13 (1989), 1317–1327.
7 KWAPISZ, M.: Some remarks on comparison theorems for discrete inequalities. J. Diff. Equ. 79 (1989) 2, 258–265.
8 SCHAEFER, H. H.: Topological vector spaces. MacMillan Co., New York 1966.
9 VULIKH, B. Z.: Introduction to the theory of partially ordered spaces. Gousud. Izd. Fiz.-Mat. Liter., Moskva 1961 (Russian).

Received February 5, 1990, revised June 22, 1990

Address: Prof. Dr. MARIAN KWAPISZ, University of Gdańsk, Department of Mathematics, ul. Wita Stwosza 57, PL-80-952 Gdańsk, Poland

ZAMM · Z. angew. Math. Mech. 71 (1991) 7/8, 295–297 Akademie Verlag

SZULC, T.

A Criterion for Nonvanishing of Determinants

Dedicated to the memory of LOTHAR COLLATZ

MSC (1980): 15A15

This paper presents a new nonsingularity criterion for an $n \times n$ complex matrix A. Based on it a new criterion is obtained for A to have all the eigenvalues with positive real parts. Then, in consequence, a new criterion is also derived, which allows a real square matrix to be a nonsingular M-matrix under weaker assumptions than those of a well known theorem of COLLATZ.

1. Preliminaries

For a positive integer n, let $V_n(C)$ be the space of column n-vectors over the complex field, and let q be an arbitrary positive integer not exceeding n. Following [13], by a *partition* π of $V_n(C)$, we mean a collection of pairwise disjoint nontrivial linear subspaces W_j, $1 \leqq j \leqq q$, of $V_n(C)$ whose direct sum is

$$V_n(C) = W_1 \dotplus W_2 \dotplus \dots \dotplus W_q .$$

For nonnegative integers $r_0 = 0 < r_1 < \dots < r_q = n$, we assume without essential loss of generality that the subspaces W_j of $V_n(C)$ are in fact explicitly given by

$$W_j = \operatorname{span} \{e_k\}_{k=r_{j-1}+1}^{r_j}, \qquad 1 \leqq j \leqq q ,$$

where e_k is the kth unit vector in $V_n(C)$, and we denote the partition π by $\pi : \{r_j\}_{j=0}^q$.

For a fixed partition π of $V_n(C)$, let φ_j be a vector norm [5] on W_j for each $1 \leqq j \leqq q$, and let Φ_π (Φ_π^A, respectively) be the collection of all q-tuples $\varphi = (\varphi_1, \dots, \varphi_q)$ ($\varphi^A = (\varphi_1, \dots, \varphi_q)$, respectively) of norms (absolute norms [1], respectively). Notice that each $\varphi \in \Phi_\pi$, viewed as a mapping

$$V_n(C) \ni x \mapsto \left| \begin{matrix} \varphi_1(P_1 x) \\ \vdots \\ \varphi_q(P_q x) \end{matrix} \right. ,$$

where P_j, $1 \leqq j \leqq q$, denotes the projection operator from $V_n(C)$ onto W_j, is a regular vectorial norm of order q on $V_n(C)$ [10].

Let A be an arbitrary $n \times n$ complex matrix representing a linear transformation from $V_n(C)$ into itself. Then, for a fixed partition $\pi : \{r_j\}_{j=0}^q$ of $V_n(C)$,

$$A = \begin{bmatrix} A_{11} & A_{12} & \dots & A_{1q} \\ A_{21} & A_{22} & \dots & A_{2q} \\ \dots & \dots & \dots & \dots \\ A_{q1} & A_{q2} & \dots & A_{qq} \end{bmatrix} \tag{1}$$

denotes the block partitioning of A with respect to π, i.e., $A_{ij} = P_i A P_j$ is a linear transformation from W_j into W_i, $1 \leqq i, j \leqq q$.

The norm $\|A_{ij}\|^\varphi$ of A_{ij} is defined as usual by

$$\|A_{ij}\|^\varphi = \sup_{\substack{x \in W_j \\ x \neq \Theta}} \left\{ \frac{\varphi_i(A_{ij}x)}{\varphi_j(x)} \right\}, \tag{2}$$

where $\varphi \in \Phi_\pi$ (in this paper Θ denotes the null vector or matrix with dimensions following from the context).

For the partitioned matrix A of (1) and for any $\varphi \in \Phi_\pi$, we define the real matrix $M^\varphi(A) = (m_{ij}^\varphi(A_{ij}))$, $1 \leqq i, j \leqq q$, by

$$m_{ij}^\varphi(A_{ij}) = \begin{cases} \inf\limits_{\substack{x \in W_i \\ x \neq \Theta}} \left\{ \dfrac{\varphi_i(A_{ii}x)}{\varphi_i(x)} \right\} & \text{if } i = j , \\ \|A_{ij}\|^\varphi & \text{if } i \neq j , \end{cases} \tag{3}$$

and we denote the sum of all off-diagonal elements of the ith row, $1 \leqq i \leqq q$, of $M^\varphi(A)$ by $P_i^\varphi(A)$. One can show [6] that

$$m_{ii}^\varphi(A_{ii}) = \begin{cases} 0 & \text{if } A_{ii} \text{ is singular} , \\ \dfrac{1}{\|A_{ii}^{-1}\|^\varphi} & \text{otherwise} . \end{cases} \tag{4}$$

Moreover, with the matrix $M^\varphi(A)$ we associate the matrix $\mathcal{T}^\varphi(A) = \operatorname{diag}(T_1^\varphi(A), \dots, T_q^\varphi(A))$ and the set

$$S(M^\varphi(A)) = \{i \in \{1, \dots, q\} : m_{ii}^\varphi(A_{ii}) > \min(P_i^\varphi(A), T_i^\varphi(A))\} ,$$

where $T_i^\varphi(A) = \max\limits_{j \in J(i)} \{m_{ji}^\varphi(A_{ji})\}$ ($J(i)$ is a set obtained from the set $J = \{1, \dots, q\}$ by removing the index i).

2. The main result

Theorem 1: *For the partitioned matrix A of* (1) *let, for a certain $\varphi \in \Phi_\pi$,*

$$m_{ii}^\varphi(A_{ii}) \geqq \min(P_i^\varphi(A), T_i^\varphi(A)) , \qquad i = 1, \dots, q , \tag{5}$$

with $\sum\limits_{j=1}^q (m_{jj}^\varphi(A_{jj}) - \min(P_j^\varphi(A), T_j^\varphi(A))) > 0$ *and*

$$m_{ii}^\varphi(A_{ii}) + m_{jj}^\varphi(A_{jj}) \geqq P_i^\varphi(A) + P_j^\varphi(A) \quad \text{for all} \quad j \in J(i) . \tag{6}$$

Furthermore, assume that for each $i \in S(M^\varphi(A))$ the inequality (6) *is strict and that for each $i \notin S(M^\varphi(A))$ there is a sequence of positive elements of $\mathcal{T}^\varphi(A)\, M^\varphi(A)$ of the form $T_i^\varphi(A)\, m_{ii_1}^\varphi(A_{ii_1})$, $T_{i_1}^\varphi(A)\, m_{i_1 i_2}^\varphi(A_{i_1 i_2}), \dots, T_{i_r}^\varphi(A)\, m_{i_r i_{r+1}}^\varphi(A_{i_r i_{r+1}})$ with $i_{r+1} \in S(M^\varphi(A))$. Then A is nonsingular.*

Proof: Applying reasoning similar to that used for unpartitioned A by PUPKOV [9], we suppose the contrary. Then there exists a nontrivial solution $x = x_1 + \dots + x_q$, $x_i \in W_i$, of the equation

$$Ax = \Theta . \tag{7}$$

The block partitioning of A implies that (7) can be rewritten as the system of the form

$$\sum_{j=1}^q A_{ij} x_j = \Theta , \qquad i = 1, \dots, q , \tag{8}$$

where $x_j = P_j x$. So, by the respective properties of vector norms [5] and by (3), we get from (8),

$$m_{ii}^\varphi(A_{ii})\, \varphi_i(x_i) \leqq \sum_{j \in J(i)} m_{ij}^\varphi(A_{ij})\, \varphi_j(x_j) , \qquad i = 1, \dots, q , \tag{9}$$

where the norms φ_i constitute φ. By the assumptions it follows that all $m_{ii}^\varphi(A_{ii})$ are positive. Therefore, in view of (9), there exists a pair of components of x, indexed by s and t, such that

$$\varphi_s(x_s) \geqq \varphi_t(x_t) \geqq \varphi_j(x_j)\,, \qquad j = 1, \ldots, s-1, s+1, \ldots, q\,, \tag{10}$$

with $\varphi_t(x_t) > 0$.

Define the set $Q^\varphi(A) = \{i \in J : \min\,(P_i^\varphi(A), T_i^\varphi(A)) = 0\}$. Assuming that $Q^\varphi(A) = \emptyset$, we shall consider the following two cases.

Case 1: $t \in S(M^\varphi(A))$. – Then, taking into account the sth inequality of (9), by (10), we get

$$m_{ss}^\varphi(A_{ss}) \leqq \frac{\varphi_t(x_t)}{\varphi_s(x_s)}\, P_s^\varphi(A) = \alpha P_s^\varphi(A)\,, \tag{11}$$

where α satisfies

$$0 < \alpha \leqq 1\,. \tag{12}$$

On the other hand, from the tth inequality of (9), we obtain

$$m_{tt}^\varphi(A_{tt}) \leqq P_t^\varphi(A) - m_{ts}^\varphi(A_{ts}) + \frac{m_{ts}^\varphi(A_{ts})}{\alpha} \leqq P_t^\varphi(A) + T_s^\varphi(A)\,\frac{(1-\alpha)}{\alpha}\,. \tag{13}$$

So, from (11) and (13) we have

$$m_{tt}^\varphi(A_{tt}) + m_{ss}^\varphi(A_{ss}) \leqq P_t^\varphi(A) + T_s^\varphi(A)\,\frac{(1-\alpha)}{\alpha} + \alpha P_s^\varphi(A)\,,$$

which, coupled with (6), becomes

$$P_s^\varphi(A) < \alpha P_s^\varphi(A) + T_s^\varphi(A)\,\frac{(1-\alpha)}{\alpha}\,. \tag{14}$$

It can be easily verified that (14) and (12) yield

$$\alpha < \min\left(1, \frac{T_s^\varphi(A)}{P_s^\varphi(A)}\right).$$

However, for such α the inequality (11) becomes

$$m_{ss}^\varphi(A_{ss}) < \min\,(P_s^\varphi(A), T_s^\varphi(A))\,,$$

which is in a contradiction to (5).

Case 2: $t \notin S(M^\varphi(A))$. – Then, by the assumptions, there is a sequence

$$T_t^\varphi(A)\, m_{ti_1}^\varphi(A_{ti_1}), \ldots, T_{i_r}^\varphi(A)\, m_{i_r i_{r+1}}^\varphi(A_{i_r i_{r+1}}) \tag{15}$$

of positive elements of $\mathcal{T}^\varphi(A)\, M^\varphi(A)$ with $i_{r+1} \in S(M^\varphi(A))$.

Keeping in mind (10), we split this part of the proof into the three following steps.

Step 1: $\varphi_t(x_t) > \varphi_{i_1}(x_{i_1})$. – Then taking into account the tth inequality of (9), (10) and the properties of the sequence (15), we obtain

$$m_{tt}^\varphi(A_{tt}) \leqq \sum_{j \in J(t)\setminus\{i_1\}} m_{tj}^\varphi(A_{tj})\,\frac{\varphi_j(x_j)}{\varphi_t(x_t)} + m_{ti_1}^\varphi(A_{ti_1})\,\frac{\varphi_{i_1}(x_{i_1})}{\varphi_t(x_t)}$$

$$< P_t^\varphi(A) - m_{ts}^\varphi(A_{ts}) + \frac{m_{ts}^\varphi(A_{ts})}{\alpha} \leqq P_t^\varphi(A) + T_s^\varphi(A)\,\frac{(1-\alpha)}{\alpha}\,.$$

On the other hand, from the sth inequality of (9) we get (11). So, the assertion follows by the corresponding part of the proof of Case 1.

Step 2: $\varphi_t(x_t) = \varphi_{i_1}(x_{i_1})$ and $i_1 \in S(M^\varphi(A))$. – The assertion follows by reasoning used in the proof of Case 1 (we take into account the pair of inequalities of (9) indexed by s and i_1).

Step 3: $\varphi_t(x_t) = \varphi_{i_1}(x_{i_1})$ and $i_1 \notin S(M^\varphi(A))$. – Then there exists an index l, $1 < l \leqq r + 1$, such that

$$\varphi_{i_k}(x_{i_k}) = \varphi_t(x_t)\,, \qquad k = 1, \ldots, l-1\,, \tag{16}$$

and either

$$\varphi_{i_l}(x_{i_l}) < \varphi_t(x_t) \tag{17}$$

or

$$\varphi_{i_l}(x_{i_l}) = \varphi_t(x_t) \quad \text{and} \quad i_l \in S(M^\varphi(A))\,. \tag{18}$$

If (17) holds then taking into account (10), (16), the properties of the sequence (15) and the i_{l-1}th inequality of (9), we obtain

$$m_{i_{l-1}i_{l-1}}^\varphi(A_{i_{l-1}i_{l-1}}) \leqq \sum_{j \in J(i_{l-1})} m_{i_{l-1}j}^\varphi(A_{i_{l-1}j})\,\frac{\varphi_j(x_j)}{\varphi_{i_{l-1}}(x_{i_{l-1}})}$$

$$= \sum_{j \in J(i_{l-1})\setminus\{i_l\}} m_{i_{l-1}j}^\varphi(A_{i_{l-1}j})\,\frac{\varphi_j(x_j)}{\varphi_{i_{l-1}}(x_{i_{l-1}})} + m_{i_{l-1}i_l}^\varphi(A_{i_{l-1}i_l})\,\frac{\varphi_{i_l}(x_{i_l})}{\varphi_{i_{l-1}}(x_{i_{l-1}})}$$

$$< P_{i_{l-1}}^\varphi(A) - m_{i_{l-1}s}^\varphi(A_{i_{l-1}s}) + \frac{m_{i_{l-1}s}^\varphi(A_{i_{l-1}s})}{\alpha}$$

$$\leqq P_{i_{l-1}}^\varphi(A) + T_s^\varphi(A)\,\frac{(1-\alpha)}{\alpha}\,.$$

On the other hand, from the sth inequality of (9) we get (11), which, coupled with (19) and (6), becomes (14). So, the assertion follows by the corresponding part of the proof of Case 1.

If (18) holds then the assertion follows by reasoning like in Case 1 (we take into account the pair of inequalities of (9) indexed by s and i_l).

In order to complete the proof we note that if $Q^\varphi(A) \neq \emptyset$ then either there exists a submatrix $\tilde{A}$ of A, formed by the blocks A_{ij} with $i, j \in J \setminus Q^\varphi(A)$, which satisfies the assumptions of the theorem or

$$\det M^\varphi(A) = \prod_{i=1}^{q} m_{ii}^\varphi(A_{ii})\,.$$

In the former case we apply, to $M^\varphi(\tilde{A})$, reasoning similar as was done above, whereas in the latter case the nonsingularity of A follows from (5) (we recall that, by the assumptions, all $m_{ii}^\varphi(A_{ii})$ are positive). So, the proof is completed. □

Remark 1: In the case when all matrices A_{ij} are 1×1 ones, Theorem 1 allows a matrix for being nonsingular under weaker assumptions than those of a theorem of PUPKOV [9] and a criterion of [11].

Remark 2: Theorem 1 allows a matrix for being nonsingular under weaker assumptions than those of Theorem 1 of [4] and Theorem 1 of [12].

3. Applications

Following OSTROWSKI [8] a real $n \times n$ matrix $C = (c_{ij})$ is called an *M-matrix* if it can be written in the form

$$C = sI - B\,, \quad s > 0,\ B \text{ is a nonnegative matrix and } \varrho(B) \leqq s\,.$$

Here ϱ denotes the spectral radius and I is the unit matrix. If $\varrho(B) < s$, then C is called a *nonsingular M*-matrix; otherwise, a *singular M*-matrix.

Theorem 2: *Let the partitioned matrix A of* (1) *be such that its diagonal blocks A_{ii} are all real, and let it satisfy, for a certain $\varphi^A \in \Phi_\pi^A$, the assumptions of Theorem* 1. *Furthermore, assume that each block A_{ii} is a nonsingular M-matrix, $1 \leqq i \leqq q$. Then all the eigenvalues of A have positive real parts.*

Proof: Let z be any complex number with $\mathrm{Re}\,(z) \leqq 0$. Then, by the reasoning used in the proof of Theorem 9 of [4] and by (4), we get

$$m_{ii}^{\varphi^A}(A_{ii} - zI) \geqq m_{ii}^{\varphi^A}(A_{ii})\,, \qquad i = 1, \ldots, q$$

(here the dimensions of the unit matrix I follow from the context). So, it can be easily verified that, for any z with $\mathrm{Re}\,(z) \leqq 0$, the matrix $A - zI$ satisfies the assumptions of Theorem 1, and therefore it is nonsingular. Thus, if λ is an eigenvalue of A, then $\mathrm{Re}\,(\lambda) > 0$, which completes the proof. □

Remark 3: Theorem 2 allows a matrix to have all the eigenvalues with positive real parts under weaker assumptions than those of Theorem 9 of [4] and Theorem 2 of [12].

Denote by $Z^{n \times n}$ the class of all real $n \times n$ matrices whose off-diagonal elements are all nonpositive.

Corollary: *Let the partitioned, as in* (1), *matrix $A \in Z^{n \times n}$ satisfy, for certain a $\varphi^A \in \Phi_\pi^A$, the assumptions of Theorem* 1, *and let each block A_{ii} be a nonsingular M-matrix, $1 \leqq i \leqq q$. Then A is a nonsingular M-matrix.*

Proof: The assertion follows by Theorem 2 and by Theorem (2, 3) [2, p. 134]. □

Remark 4: In the special case when all the matrices A_{ij} are 1×1 ones, Corollary allows $A \in Z^{n \times n}$ to be a nonsingular M-matrix under weaker assumptions than those of Theorem 2.4.14 of [7] (see also in [3]).

References

1 BAUER, F. L.; STOER, J.; WITZGALL, W.: Absolute and monotonic norms. Numer. Math. 3 (1961), 257–264.
2 BERMAN, A.; PLEMMONS, R. J.: Nonnegative matrices in the mathematical sciences. Academic Press, New York 1979.
3 COLLATZ, L.: Aufgaben monotoner Art. Arch. der Math. 3 (1952), 366–376.
4 FEINGOLD, D. G.; VARGA, R: S.: Block diagonally dominant matrices and generalizations of the Gerschgorin circle theorem. Pacific J. Math. 12 (1962), 1241–1250.
5 HOUSEHOLDER, A.: The theory of matrices in numerical analysis. Blaisdell, New York 1964.
6 LANCASTER, P.: Theory of matrices (in Russian). Nauka, Moscow 1976.
7 ORTEGA, J. M.; RHEINBOLDT, W. C.: Iterative solution of nonlinear equations in several variables (in Russian). Mir, Moscow 1975.
8 OSTROWSKI, A. M.: Über die Determinanten mit überwiegender Hauptdiagonale. Comment. Math. Helvet. 10 (1937), 69–96.
9 PUPKOV, V. A. (Пупков, В. А.) Об изолированном собственном значении матрицы и структуре его собственного вектора. Журн. вычислит. мат. и мат. физ. 23 (1983), 1304–1313.
10 ROBERT, F.: Blocs-H-matrices et convergence des méthodes itératives classiques par blocs. Linear Algebra Appl. 2 (1969), 223–265.
11 SHIVAKUMAR, P. N.; CHEW, K. H.: A sufficient condition for nonvanishing of determinants. Proc. Amer. Math. Soc. 43 (1974) 1, 63–66.
12 SZULC, T.: A nonsingularity criterion for complex matrices. Linear Algebra Appl. (submitted).
13 VARGA, R. S.: Minimal Gerschgorin sets for partitioned matrices. SIAM J. Numer. Anal. 7 (1970), 493–507.

Received May 16, 1990, revised August 30, 1990

Address: Dr. TOMASZ SZULC, Instytut Matematyki UAM, ul. Matejki 48/49, PL-60-769 Poznań, Poland

ZAMM · Z. angew. Math. Mech. 71 (1991) 7/8, 297–298 Akademie Verlag

GRIEPENTROG, E.

The Numerical Treatment of the Motion Equations of Mass Point Systems

Dedicated to the memory of LOTHAR COLLATZ

MSC (1980): 65C99

A new method for the transformation of the motion equations of mass point systems into an explicit ordinary differential equation system is proposed. Using a simple projection process the approximate solution produced by a numerical integration method is forced to proceed inside the constraint manifold.

We consider the motion equations of mass point systems

$$M\dot{v} = f(v, r, t) - G^T(r, t)\, z\,, \qquad \dot{r} = v\,, \qquad 0 = g(r, t)\,, \tag{1}$$

with $G(r, t) := [\partial g/\partial r]\,(r, t)$. Here $G^T(r, t)$ represents the transpose of the Jacobian matrix $G(r, t)$, $r \in R^{3n}$ the n mass point positions, v the corresponding velocities and z the Lagrange factors. For simplicity we assume a constant mass matrix M; the case $M = M(r)$ can be treated analogously, but the arising formulas are much more complicated.

It is well-known (cf. [1, 2, 5]) that the differential algebraic equation (DAE) (1) has index 3, and that the numerical integration of index-3-problems presents very serious difficulties. Therefore methods for index reduction are used to get lower index equations. If those methods are based on the repeated differentiation of the constraint equations

$$g(r, t) = 0\,, \qquad g \in R^k\,, \tag{2}$$

in the produced lower index systems stability problems arise [1].

In [3] we suggested methods for index reduction via integration of the variables whose derivatives do not occur, in order to receive the well-posed index-1-problems (cf. [4]).

The application of this concept proposes the introduction of the new variable vector

$$u = v + M^{-1}G^T(r, t) \int_0^t z(s)\,\mathrm{d}s\,. \tag{3}$$

With $w(t) = \int_0^t z(s)\,\mathrm{d}s$ we then obtain the equation

$$M\dot{v} = M\dot{u} - ([\partial G^T/\partial r]\, v + \partial G^T/\partial t)\, w - G^T\dot{w}\,,$$

i.e.

$$\begin{aligned} M\dot{u} &= f(u - M^{-1}G^Tw, r, t) \\ &\quad + \{[\partial G^T/\partial r](u - M^{-1}G^Tw) + \partial G^T/\partial t\}\, w\,, \\ \dot{r} &= u - M^{-1}G^Tw\,. \end{aligned} \tag{4}$$

In velocity description the equations $g(r, t) = 0$ yield

$$Gu - GM^{-1}G^Tw + \partial g/\partial t = 0\,, \tag{5}$$

where the matrix $GM^{-1}G^T$ is nonsingular for linearly independent constraint equations. Thus we obtain

$$w = (GM^{-1}G^T)^{-1}\,(Gu + \partial g/\partial t)\,, \qquad u - M^{-1}G^Tw = Qu - q$$

with

$$Q := I - M^{-1}G^T(GM^{-1}G^T)^{-1}\, G\,,$$

$$q := M^{-1}G^T(GM^{-1}G^T)^{-1}\, \partial g/\partial t\,.$$

$Q = Q(r, t)$ is the projector onto the tangential space $\ker(G)$ of the constraint manifold along $M^{-1}\,\mathrm{im}\,(G^T)$.

Hence we receive the explicit ordinary differential equation system

$$\left.\begin{aligned} M\dot{u} &= f(Qu - q, r, t) + \{[\partial G^T/\partial r]\,[Qu - q] + \partial G^T/\partial t\}\, w\,, \\ \dot{r} &= Qu - q \end{aligned}\right\} \tag{6}$$

with $w = (GM^{-1}G^T)^{-1}\,(Gu + \partial g/\partial t)$ and the restrictions $g(r(0), 0) = 0$, $G(r(0), 0)\, u(0) = -\,[\partial g/\partial t]\,(r(0), 0)$ for the initial vectors $r(0)$ and $u(0)$. The functions $v(t)$ and $z(t)$ may be calculated using the formulas

$$v = Qu - q\,, \qquad z = \mathrm{d}([GM^{-1}G^T]^{-1}\,[Gu + \partial g/\partial t])/\mathrm{d}t\,. \tag{7}$$

If the constraints are independent of the time t, then the equations (4) reduce to

$$M\dot{u} = f(Qu, r, t) + [\partial G^T/\partial r]\, vw\,, \qquad \dot{r} = Qu \tag{8}$$

with $v = \dot{r}$, $w = [G(r)\, M^{-1}G^T(r)]^{-1}\, G(r)\, u$.

The calculation of $[\partial G^T/\partial r]\, vw$ takes place by the formula

$$[\partial G^T/\partial r]\, vw = \sum_{m=1}^{k} w_m H_m(r)\, v \tag{9}$$

using the Hesse-matrices $H_m(r)$ of the components $g_m(r)$ of $g(r)$. Consequently, we have to solve

$$M\dot{u} = f(v, r, t) + \sum_{m=1}^{k} w_m H_m(r)\, v\,, \qquad \dot{r} = v \tag{10}$$

using

$$V(r) := [G(r)\, M^{-1}G^T(r)]^{-1}, \; w = V(r)\, G(r)\, u, \; v = u - M^{-1}G^T(r)\, w\,.$$

If the restrictions $g_m(r) = 0$ express rigid connections between the mass points, the functions $g_m(r)$ are quadratic. Hence the elements of the matrices $G(r)$ can be calculated exactly as arbitrary symmetric difference quotients

$$[\partial g_m/\partial r_j]\,(r) = [g_m(r + he_j) - g_m(r - he_j)]/h\,.$$

Of course it is possible to calculate $\partial g_m/\partial r_j$ algebraically, but using a computer the suggested way is more convenient since it reduces the user's programming effort for his particular task. Moreover, the Hesse-matrices $H_m(r)$ are constant in this case.

Now we return to the general situation. Since $Q(r)$ is a projector onto the tangential space $TS(r) := \ker(G(r))$ of the constraint manifold S defined by the equation $g(r) = 0$, the differential equation

$$\dot{r} = Q(r)\, u(t) \tag{11}$$

implies for each function $u(t)$ that $r(t)$ remains on S. Namely, (11) yields $[I - Q(r)]\,\dot{r} = 0$, i.e. $\mathrm{d}[g(r(t))]/\mathrm{d}t = G(r)\,\dot{r} = 0$. Therefore the approximate solution vector r_j for $r(t_j)$, which is produced by a numerical integration method with the order p of convergence, leads to an error $g(r_j) = O(h^p)$ (h stepsize) at most.

On the other hand, the modification of the r_j by vectors z_j of the magnitude $O(h^{p+1})$ does not change the order of convergence. Therefore it is reasonable to add vectors z_j ensuring

$$g(r_j + z_j) = 0 \tag{12}$$

with a very high accuracy. Clearly, the correction z_j should have no component in the tangential space TS (r_j), i.e. it satisfies $Q(r_j)\, z_j = 0$ suggesting $z_j = M^{-1} G^{\mathrm{T}}(r_j)\, s_j$ with $s_j \in R^k$.

Newton's method applied to the equation

$$g(r + M^{-1} G^{\mathrm{T}}(r)\, s) = 0 \tag{13}$$

yields in the first step $G(r)\, M^{-1} G^{\mathrm{T}}(r)\, s_1 = -g(r)$, i.e.

$$s_1 = -V(r)\, g(r)\,.$$

Since $g(r_j) = O(h^{p+1})$ for $g(r_{j-1}) = O(h^{2p+1})$ holds, already the first Newton step guarantees for

$$\hat{r}_j := r_j + M^{-1} G^{\mathrm{T}}(r_j)\, s \tag{14}$$

the estimation $g(\hat{r}_j) = O(h^{2p+1})$ realizing the desired high accuracy for the validity of the constraint equations. Here we assume $G(r)$ and $V(r)$ to be calculated exactly; but obviously the estimation remains valid if the error of these matrices is of the order $O(h^{2p+1})$, too.

As an example we consider the equations of the plane mathematical pendulum omitting the third components $x_3 = v_3 = 0$:

$$\left.\begin{aligned} m\dot{v}_1 &= -2x_1 z\,, \qquad m\dot{v}_2 = -2x_2 z - mg\,,\\ \dot{x}_1 &= v_1\,, \qquad \dot{x}_2 = v_2\,,\\ x_1^2 &+ x_2^2 - l^2 = 0 \end{aligned}\right\} \tag{15}$$

(l length, g gravitational constant). Here we obtain $q = 0$, $V(r)^{-1} = (x_1^2 + x_2^2)^{-1}\, m/4$,

$$Q(r) = [x_1^2 + x_2^2]^{-1} \begin{pmatrix} x_2 x_2 & -x_1 x_2 \\ -x_1 x_2 & x_1 x_1 \end{pmatrix}, \quad H = \begin{pmatrix} 2 & 0 \\ 0 & 2 \end{pmatrix},$$

$$2w = m[x_1 u_1 + x_2 u_2]\,[x_1^2 + x_2^2]^{-1}$$

and finally

$$\left.\begin{aligned} \dot{u}_1 &= x_2[x_1 u_1 + x_2 u_2]\,[x_2 u_1 - x_1 u_2]\,[x_1^2 + x_2^2]^{-2}\,,\\ \dot{u}_2 &= -x_1[x_1 u_1 + x_2 u_2]\,[x_2 u_1 - x_1 u_2]\,[x_1^2 + x_2^2]^{-2} - g\,,\\ \dot{x}_1 &= x_2[x_2 u_1 - x_1 u_2]\,[x_1^2 + x_2^2]^{-1}\,,\\ \dot{x}_2 &= -x_1[x_2 u_1 - x_1 u_2]\,[x_1^2 + x_2^2]^{-1}\,. \end{aligned}\right\} \tag{16}$$

The proposed method was tested by a general program for the calculation of the plane motion of multi-mass pendula. In this case the vector $f(v, r, t)$ has the special form

$$f = (0, -m_1 g, 0, -m_2 g, \ldots, 0, -m_n g)^{\mathrm{T}}\,,$$

and the constraints are quadratic equations. The right-hand sides of the ordinary differential equation system

$$M\dot{u} = f - \left(\sum_{m=1}^{k} w_m H_m\right) Qu\,, \qquad \dot{r} = Qu\,, \tag{17}$$

are calculated in the way proposed in (10). For the numerical integration a simple RK4-method without stepsize control was used, but each complete integration step was followed by a correction according to (14). For some comparison calculations the correction was omitted.

In the case of the single mass pendulum we considered the initial swing angle $\pi/2$ (from the perpendicular line). By the choice $g = 9.80665\ \mathrm{m/sec^2}$ and $l = 0.71319163293$ m the period of 2 sec was attained. The following table shows the deviation of r from the constraint manifold and the error $90°\text{-}\alpha$ of the computed angle α after n periods. For all calculations the single precision arithmetics was used. The stepsizes $h = 0.01, 0.02, 0.04, 0.08$ produce 200, 100, 50, 25 integration steps per period respectively, the asterix denotes calculations without correction steps according to (14).

Table of Δr (meter) and $\Delta\alpha$ (degree)

n	$h = 0.01$	$h = 0.02$	$h = 0.02$*	$h = 0.04$	$h = 0.04$*	$h = 0.08$
5	0	0	2.9E-6	0	9.5E-4	0
	1.4E-5	5.2E-4	5.9E-4	1.7E-2	1.9E-2	6.OE-1
10	0	0	5.8E-6	0	1.9E-3	0
	2.7E-5	1.0E-3	1.2E-3	3.4E-2	3.8E-2	1.68
15	0	0	8.8E-6	0	2.8E-3	0
	4.1E-5	1.5E-3	1.7E-3	5.3E-2	5.8E-2	4.77
20	0	0	1.2E-5	0	3.8E-3	0
	5.5E-5	2.0E-3	2.3E-3	7.5E-2	7.8E-2	12.5

Obviously the correction procedure does not deliver an essentially higher precision of the computed solution (characterized by the angle α), but it guarantees for sufficiently small stepsizes that r remains on the constraint manifold.

As a second example we considered the double pendulum with the masses $m_1 = m_2 = 1$ and the constraint equations

$$x_1^2 + y_1^2 - 0.25 = 0\,, \qquad (x_2 - x_1)^2 + (y_2 - y_1)^2 - 0.09 = 0$$

for the initial constellation $x_1 = 0.5$, $y_1 = 0$, $x_2 = 0.2$, $y_2 = 0$; $\dot{x}_1 = \dot{x}_2 = \dot{y}_1 = \dot{y}_2 = 0$. Using the stepsize 0.01 sec also in this case the correction steps (14) secured that the solution could not drift away from the constraint manifold.

References

1 Führer, C.: Lösen von mechanischen Bewegungsgleichungen mit Zwangsbedingungen. ZAMM **69** (1989) 4, T 32–T 35.

2 Gear, C. W.; Gupta, G. K.; Leimkuhler, B.: Automatic integration of Euler-Lagrange equations with constraints. J. Comput. Appl. Math. **12, 13** (1985), 77–90.

3 Griepentrog, E.: Transferability of higher-index differential-algebraic equations. Univ. Greifswald (DDR), Preprint-Reihe Mathematik 1988, Nr. 20, pp. 3–19.

4 Griepentrog, E.; März, R.: Differential-algebraic equations and their numerical treatment. Teubner, Leipzig 1986.

5 Hairer, E.; Lubich, C.; Roche, M.: The numerical solution of differential-algebraic systems by Runge-Kutta methods. Lecture notes in Mathematics 1409, Berlin-Heidelberg, Springer-V. 1989.

Received February 19, 1990, revised March 20, 1990

Address: Prof. Dr. sc. Eberhard Griepentrog, Abteilung Math./Naturwiss. der Ingenieurhochschule Berlin, O-1134 Berlin, Germany

ZAMM · Z. angew. Math. Mech. 71 (1991) 7/8, 298–300 Akademie Verlag

Zorin, I. S.; Morozov, N. F.

On the Theory of Quasiequilibrium Cracks

Dedicated to the memory of Lothar Collatz

MSC (1980): 73M05

The equilibrium cracks theory, proposed by V. V. Novozhilov in [1, 2], holds that the load level P is consistent with the equilibrium crack length interval $[l_-(P), l_+(P)]$. According to A. Griffith and G. Irwin theory this interval is equal to one point. V. V. Novozhilov suggests the structural load independent parameter

$$\varkappa = l_+(P)/l_-(P)\,. \tag{1}$$

R. Thomson, C. Hsieh, V. Rava [3] and R. Thomson [4] find analogous results for discrete models of continuous media. For the equilibrium cracks with length $2l$ they hold the possible load level interval $[P_-(l), P_+(l)]$ and introduce the length independent parameter

$$R = P_+(l)/P_-(l)\,. \tag{2}$$

The grating capture R is known in the literature as measure of a result reliable of numerical modelling in deformation problems of rigid body mechanics.

Note that a simple algebraic dependence between $\varkappa$ and R values was obtained in [5]. For the general modelling crack problem in an elastic plane with loading by a normal force at infinity we can find that $R = 1/\sqrt{\varkappa}$. Because of $R = (U_*^2 + 1)/2U^*$, $U_*^2 = \pi\beta/4\alpha$, $\beta = \gamma E/D\sigma_c^2$ and taking into account that α is a known function of β we obtain γ as a function of R. An experimental value of R may be received by statistic analysis of equilibrium cracks in a monocrystal.

The article devotes to investigation the dynamic analogues of R and $\varkappa$ for a quasiequilibrium crack. The last moves in the elastic body with a constant speed and there remains the own length.

In the first place we consider the general problem and give an account away of the investigation (see [6]).

The crack with length $2l$ moves in the O_x direction with a constant speed v. Suppose that the concentrated split forces P and the couple forces σ_c, distributed in the Δ_- and Δ_+-neighbourhoods of the crack tips, acts to its shores. Choose the beginning of the moving coordinate system $(\xi, \eta) = (x_1 - vt, x_2)$ in the centre of the crack. Then the loads offers by the vector

$$P = \{p_1, p_2\} = \{O, \delta(L)\,p + [H(-l) - H(l) + H(l - \Delta_+) - H(-l + \Delta_-)]\,\sigma_c\} \tag{3}$$

Here δ is the Dirac function, H the Heaviside function, L the distance between the points P and O.

The solution of the Lamé equations is constructed by means of the potentials $\varphi_j(\xi, \eta)$, $j = 1, 2$ and

$$U_1 = \frac{\partial\varphi_1}{\partial\xi} + \frac{\partial\varphi_2}{\partial\xi}, \qquad U_2 = \frac{\partial\varphi_1}{\partial\xi} - \frac{\partial\varphi_2}{\partial\eta}. \tag{4}$$

The functions φ_j, $j = 1, 2$ are the solutions of the equations

$$\alpha_j^2 \frac{\partial^2\varphi_j}{\partial\xi^2} + \frac{\partial^2\varphi_j}{\partial\eta^2} = O\,. \tag{5}$$

Here $\alpha_j^2 = 1 - v^2/c_j^2$, $c_1^2 = (\lambda + 2\mu)/\varrho$, $c_2^2 = \mu/\varrho$.

Acording to the L. A. Galin [7] method, consider the complex potentials

$$\omega_j(z_j) = \varphi_j(z_j) + i\psi_j(z_j)\,, \qquad j = 1, 2\,. \tag{6}$$

Then the boundary conditions on the crack surface are transformed to the kind

$$(1 + \alpha_2^2)\operatorname{Re}\Omega_2 - 2\alpha_1 \operatorname{Im}\Omega_1 = O\,,$$
$$-2\alpha_2 \operatorname{Im}\Omega_2 + (1 + \alpha_2^2)\operatorname{Re}\Omega_1 = \frac{1}{\mu}P_2\,, \tag{7}$$
$$|\xi| \leqq l\,, \qquad \eta = O\,.$$

The vector function $\Omega(z) = d^2\omega(z)/dz^2$ is obtained by the vector analogous of the M. V. Keldvish – L. I. Sedov function [8]

$$\Omega(z) = \frac{\bar a}{\pi\sqrt{z^2 - l^2}} \int_{-l}^{l} \frac{p_2\sqrt{l^2 - t^2}}{t - z}\,dt\,. \tag{8}$$

The tensor of stresses is recovered by means of Ω_1 and Ω_2 and thus for $\xi >$

$$\sigma_{22}(\xi, O) = \frac{1}{\pi}\left\{\sigma_e\left[\arccos\frac{l^2 + \xi(l - \Delta_-)}{l(l + \xi - \Delta_-)} + \arccos\frac{l^2 - \xi(l - \Delta_+)}{l(l - \xi - \Delta_+)} + \frac{\sqrt{\Delta_-(2l - \Delta_-)} - \sqrt{\Delta_+(2l - \Delta_+)}}{\sqrt{\zeta^2 - l^2}} + \frac{-\xi(\arccos(1 - \Delta_-/l) + \arccos(1 - \Delta_+/l))}{\sqrt{\xi^2 - l^2}}\right] - \frac{P\sqrt{l^2 - L^2}}{(L - \xi)\sqrt{\xi^2 - l^2}}\right\}. \tag{9}$$

The formula for $\sigma_{22}(\xi, O)$ when $\xi < -l$ is obtained on the analogous way.

For the normal component of the crack displacement vector we obtain

$$U_2(\xi + O) = \frac{\alpha_1(1 - \alpha_2^2)}{\pi\mu r(v)} \times \left\{p \ln\left|\frac{\sqrt{l^2 - L^2}\sqrt{l^2 - \xi^2} + l^2 - L\xi}{l(L - \xi)}\right| - \sigma_c\left[(\arccos(1 - \Delta_-/l) + \arccos(1 - \Delta_+/l))\sqrt{l^2 - \xi^2} + (\xi - l + \Delta_+)\ln\left|\frac{\sqrt{(2l - \Delta_+)\Delta_+}\sqrt{l^2 - \xi^2} + l^2 - \xi(l - \Delta_+)}{l(l - \xi - \Delta_+)}\right| - (\xi + l - \Delta_-) \times \ln\left|\frac{\sqrt{(2l - \Delta_-)\Delta_-}\sqrt{l^2 - \xi^2} + l^2 + \xi(l - \Delta_-)}{l(l + \xi - \Delta_-)}\right|\right]\right\}. \tag{10}$$

Note that when P and L are given, the values of l, Δ_+, Δ_- and v are unknown in the problem of quasiequilibrium crack. As already mentioned it is necessary to lay the restrictions as well as in the static case.

$$\frac{1}{D}\int_{l}^{l+D} \sigma_{22}(\xi, O)\,d\xi \geqq \sigma_c\,, \qquad \frac{1}{D}\int_{-l-D}^{-l} \sigma_{22}(\xi, 0)\,d\xi \leqq \sigma_c\,. \tag{11}$$

Now we shall study concrete examples to take the results and the static crack problems data.

Consider the elastic plane with quasiequilibrium crack on the plane stress condition. The loads are the even functions of ξ, η coordinates. By force of symmetry of the problem the unknown kinematical parameters are in the connection with the equation which is analogous to the equation obtained in [2].

For example, the crack is free of the stresses and the loads P acts at infinity. The normal shore displacement is the following

$$U_2(\xi, +O) = \frac{\alpha_1(1 - \alpha_2^2)}{\mu r(v)}\left\{p\sqrt{l^2 - \xi^2} + \frac{\sigma_c}{2}\left[(\xi - l + \Delta) \times \ln\frac{l^2 - \xi(l - \Delta) - \sqrt{\Delta(2l - \Delta)}\sqrt{l^2 - \xi^2}}{l^2 - \xi(l - \Delta) + \sqrt{\Delta(2l - \Delta)}\sqrt{l^2 - \xi^2}} - (\xi + l - \Delta)\ln\frac{l^2 + \xi(l - \Delta) - \sqrt{\Delta(2l - \Delta)}\sqrt{l^2 - \xi^2}}{l^2 + \xi(l - \Delta) + \sqrt{\Delta(2l - \Delta)}\sqrt{l^2 - \xi^2}} - 4\sqrt{l^2 - \xi^2}(\arccos(1 - \Delta/l))\right]\right\}, \qquad |\xi| < l\,. \tag{12}$$

According to the even property of $U_2(\xi)$ and the opposite property of $dU_2(\xi)/d\xi$, the clearly conditions $dU_2/dt > O$ and $dU_2/dt < O$ on the right and left neighbourhoods of the crack tips are confirmed.

Taking into account the stress tensor and constructing the necessary calculations, we obtain the equation for

$$\sqrt{\alpha}\,(1+\alpha)\left(1-\frac{2}{\pi}\operatorname{arcctg}\frac{1}{\sqrt{\alpha}}\right)=\beta(v)\,, \tag{13}$$

with

$$\beta(v)=E\gamma'(v)/D\sigma_c^2\,.$$

The forms of this equation and the equation from [2] are identical. The speed dependence is in the right part of (13).

$$\left.\begin{aligned}\gamma'(v)&=\gamma\,\frac{(1+\nu)\,\alpha_1(v)\,(1-\alpha_2^2(v))}{r(v)}\,,\\ r(v)&=4\alpha_1(v)\,\alpha_2(v)-(1-\alpha_2^2(v))^2\,.\end{aligned}\right\} \tag{14}$$

The fact that the static specific density of surface energy γ is the limiting value of γ' is proved by direct calculations.

The $\varkappa$ value may be obtained thus from formula [2] with substitution γ' to γ. The dependence $\varkappa=\varkappa(v)$ is shown in Fig. 1.

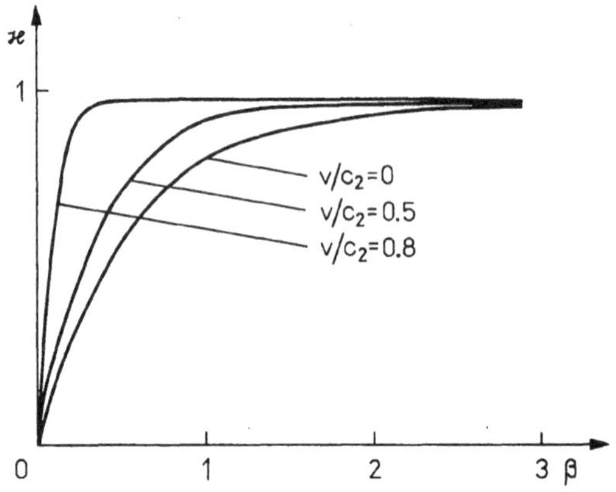

Fig. 1

In conclusion we consider the case when the crack moving is called by the mechanical captures acting for the crack shores with density P on the interval $[-L-\Delta,\,-L+\Delta]$ of the Ox axis.

The analytical solution of the problem is constructed as pointed out previously. In this case it is interesting to investigate the dependence between the crack speed propagation and the capture moving speed. The normal component of the crack shores speed is calculated from the expression

$$\frac{\partial U_2}{\partial t}\,(\xi,\pm O)=v(\sqrt{1-v^2/c_1^2}\,\operatorname{Im}\Omega_1+\operatorname{Re}\Omega_2)\qquad(\xi,\pm O)\,.$$

By means of (9) we obtain

$$\begin{aligned}\frac{\partial U_2}{\partial t}\,(\xi,+O)&=-\,\frac{\sqrt{1-v^2/c_1^2}\,v^3/c_2^2}{\pi\mu r(v)}\\ &\times\left\{P\left[\frac{2}{\sqrt{-\xi}}\,(\sqrt{L-\Delta}-\sqrt{L+\Delta})\right.\right.\\ &+\ln\left|\frac{(\sqrt{-\xi}+\sqrt{L+\Delta})\,(\sqrt{-\xi}-\sqrt{L-\Delta})}{(\sqrt{-\xi}-\sqrt{L+\Delta})\,(\sqrt{-\xi}+\sqrt{L+\Delta})}\right|\\ &\left.+\,\sigma_c\left[2\sqrt{-\frac{\Delta}{\xi}}-\ln\left|\frac{\sqrt{-\xi}+\sqrt{\Delta}}{\sqrt{-\xi}-\sqrt{\Delta}}\right|\right]\right\}.\end{aligned} \tag{15}$$

Suppose that the captures are far from the crack tip and its work region is fixed. Taking into consideration, as emphasized above, that the couple stresses act near the crack tip, we obtain from (16) an asymptotical formula for the capture speed

$$v_* \sim \frac{\sqrt{1-v^2/c_1^2}\,v^3/c_2^2}{2\pi\mu r(v)}\,P_*\,. \tag{16}$$

The diagram of the dependence $v=v(v_*)$ is shown on the Fig. 2.

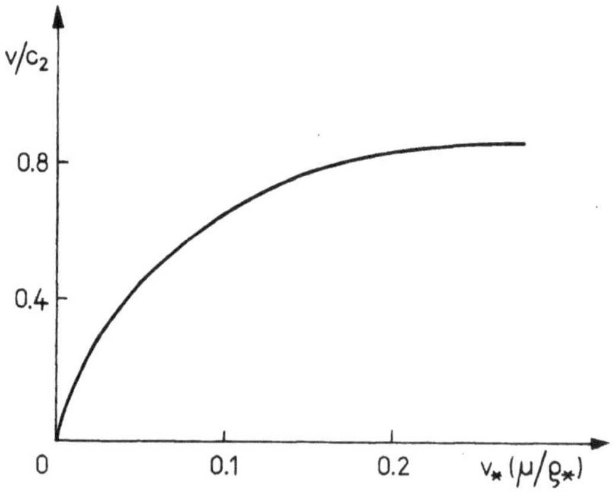

Fig. 2 (ϱ means P)

References

1 Novozhilov, V. V.: PMM, **33** (1969) 2, 212–222.
2 Novozhilov, V. V.: PMM, **33** (1969) 5, 797–812.
3 Thomson, R. M.; Hsieh, C.; Rana, V.: J. Appl. Phys. **42** (1971) 8, 3154–3160.
4 Thomson, R. M.: Proc of NATO adv. research Inst. of atom. of fracture. New York 1983, 167–204.
5 Morozov, N. F.; Paukshto, M. V.: Dokl. Acad. Sci. USSR **299** (1988) 2, 323–325.
6 Zorin, I. S.; Morozov, N. F.; Paukshto, M. V.: Proc. the 4th USSR conf. Mixture probl. mech. deform. rigid body. Odessa 1989, 142.
7 Galin, L. A.: The theory of elasticity contact problems. Nauka, Moskwa 1953, 304p.
8 Sedov, L. I.: Mechanics of continuum media. V. 2, Nauka, Moskwa 1984, 576p.

Received July 17, 1990

Address: Dr. I. S. Zorin; Prof. Dr. N. F. Morozov, 10 Krasnoarmeiskaja 4–19, Leningrad 198103, UdSSR

ZAMM · Z. angew. Math. Mech. 71 (1991) 7/8, 300–303 Akademie Verlag

Wallaschek, J.

Zur Dämpfung winderregter Schwingungen in den Bündelleitern elektrischer Freileitungen

Herrn Prof. Dr. P. Hagedorn zum 50. Geburtstag gewidmet

1. Einleitung

Bei der Übertragung von Wechselstrom kommt es zur Stromverdrängung in die Randzonen des elektrischen Leiters (Skin- oder Hauteffekt). Deshalb ordnet man die Leiterseile elektrischer Freileitungen meist in Bündeln an. Aus mechanischer Sicht ist dies recht problematisch, denn neben den klassischen *wirbelerregten Schwingungen* („Karman-Schwingungen") kommt es in Bündelleitern auch zum Auftreten von *Flatterschwingungen* der windabwärts liegenden Leiterseile („wake galloping"). Während das Problem des wake galloping bereits früh untersucht wurde und mittlerweile verschiedene Methoden zur Unterdrückung dieser Schwingungen bekannt sind [1, 2], finden die wirbelerregten Schwingungen erst in letzter Zeit zunehmendes Interesse. Dabei steht insbesondere die Entwicklung geeigneter Dämpfungsmaßnahmen im Vordergrund [3], denn langfristig können diese Schwingungen zu Materialermüdung führen.

Abstandshalter werden in mehr oder weniger regelmäßigen Abständen zwischen den Leitern eines Bündels eingebaut um eine Berührung der Seile, die sich infolge der Stromleitung anziehen, im Betrieb zu verhindern und um die Leiterseile nach hohen Kurzschlußströmen (bis ca. 100 kA) wieder zu trennen. Anfangs wurden die Abstandhalter starr ausgeführt. Diese Lösung erwies sich aber schnell als ungünstig, denn es kam häufig zur Bildung von „Schwingungsinseln" innerhalb der Teilfelder [4] und eine Bedämpfung dieser Schwingungen vom Rand her (z. B. mit Dämpfern Typ Stockbridge) erwies sich als unmöglich: die laufenden Wellen werden am Abstandhalter reflektiert und können nicht den Spannfeldrand erreichen. Aus diesem Grund wurden *selbstdämpfende Feldabstandhalter* entwickelt, die eine gewisse Elastizität, verbunden mit einer günstigen Dämpfung, aufweisen. Ein gut abgestimmter Abstandhalter soll einen möglichst großen Anteil der Schwingungsenergie der Leiterseile dissipieren.

Die Auslegung der selbstdämpfenden Abstandhalter erfolgte bisher empirisch, und es sind noch keine allgemein gültigen Konstruktionsregeln bekannt, die auf theoretischen Untersuchungen basieren. In dieser Arbeit wird eine solche Optimierungsvorschrift formuliert, indem eine quadratische Form, die die Energiedissipation des Systems beschreibt, hergeleitet wird. Die „optimalen" Systemparameter können dann anhand der Eigenwerte der zugehörigen Dissipationsmatrix bestimmt werden.

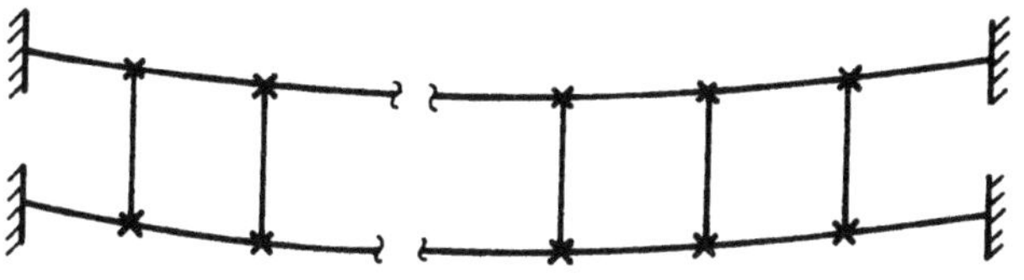

Bild 1

2. Das mechanische Modell

Wir untersuchen zunächst den Fall harmonischer Schwingungen eines senkrechten Zweierbündels. Die Verallgemeinerung der Ergebnisse auf Bündelleiter mit mehr als zwei Leiterseilen und auf horizontale Bündel ist später leicht möglich. Das Zweierbündel besteht aus zwei Teilleitern, die durch mehrere Abstandhalter miteinander verbunden sind (Bild 1). Für unsere Überlegungen genügt es, nur einen einzigen Abstandhalter und die unmittelbar benachbarten Leiterseilabschnitte zu untersuchen. In Bild 2 bezeichnet $w_i(x, t)$ die Vertikalverschiebung des Leiterseils in Abschnitt i. Die Schwingungen des Systems werden mit hinreichender Genauigkeit durch die partiellen Differentialgleichungen

$$EI_i \frac{\partial^4}{\partial x^4} w_i(x, t) - T_i \frac{\partial^2}{\partial x^2} w_i(x, t) + \mu_i \frac{\partial^2}{\partial t^2} \ddot{w}(x, t) = 0\,, \qquad (1)$$

$i = 1, 2, 3, 4$,

einer Saite (Spannkraft T, Massendichte μ) mit Biegefestigkeit EI beschrieben. Die charakteristiche Länge $\sqrt{EI/T}$ beträgt dabei nur wenige cm [5].

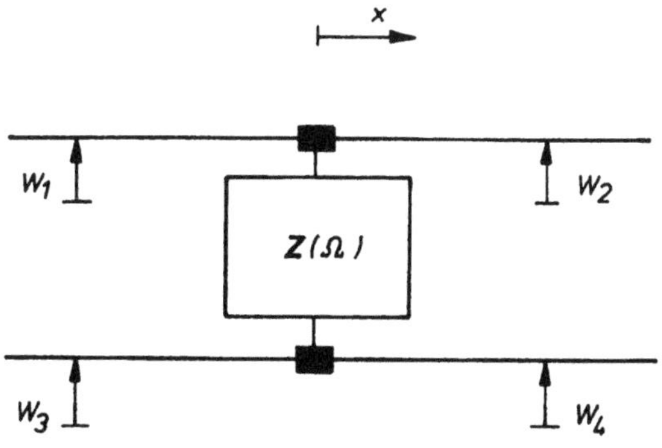

Bild 2

An der Stelle $x = 0$ ergeben sich Übergangsbedingungen, die die vier Gleichungen (1) koppeln. Die Dynamik des Abstandhalters wird dabei linearisiert durch seine 4×4 (komplexe) Impedanzmatrix $\boldsymbol{Z}(\Omega)$ beschrieben. Bei harmonischen Schwingungen gilt dann der Zusammenhang

$$[\hat{F}_1, \hat{F}_2, \hat{F}_3, \hat{F}_4]^T = \hat{\boldsymbol{F}} = \boldsymbol{Z}(\Omega)\,\hat{\boldsymbol{q}} = \boldsymbol{Z}(\Omega)\,[\hat{q}_1, \hat{q}_2, \hat{q}_3, \hat{q}_4]^T \qquad (2)$$

zwischen den Amplituden $\hat{F}_i$ der Kräfte,

$$F_i(t) = \mathrm{Re}\,[\hat{F}_i\, e^{j\Omega t}]\,, \qquad i = 1, 2, 3, 4\,, \qquad (3)$$

und den Amplituden der Geschwindigkeiten

$$\frac{\partial}{\partial t} w_i(x, t)|_{x=0} = \dot{q}_i(t) = \mathrm{Re}\,[\hat{\dot{q}}_i\, e^{j\Omega t}]\,, \qquad i = 1, 2\,,$$
$$\frac{\partial^2}{\partial x\, \partial t} w_i(x, t)|_{x=0} = \dot{q}_i(t) = \mathrm{Re}\,[\hat{\dot{q}}_i\, e^{j\Omega t}]\,, \qquad i = 3, 4\,, \qquad (4)$$

siehe Bild 3.

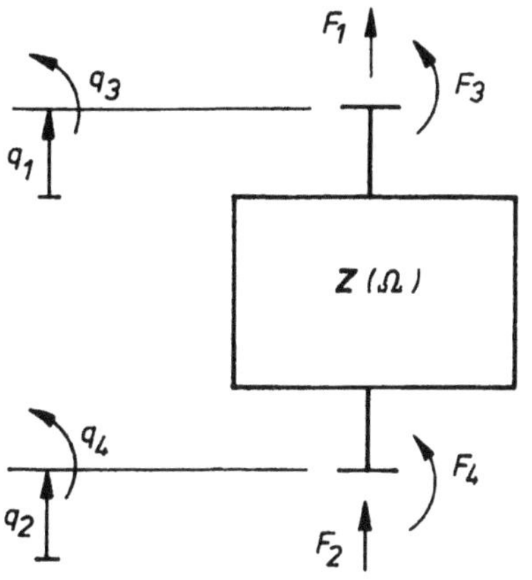

Bild 3

3. Berechnung der Leiterseilschwingungen

Für harmonische Schwingungen mit Kreisfrequenz Ω können die Lösungen von (1) in der Form

$$w_i(x, t) = \mathrm{Re}\,\{[W_i^+\, e^{-jk_i x} + W_i^-\, e^{jk_i x} + V_i^+\, e^{l_i x} + V_i^-\, e^{-l_i x}]\, e^{j\Omega t}\} \qquad (5)$$

angeschrieben werden, mit

$$\left.\begin{aligned} k_i &= \varkappa_i \left[\left[1 + \frac{\beta_i^4}{\varkappa_i^4}\right]^{1/2} - 1\right]^{1/2}, \\ l_i &= \varkappa_i \left[\left[1 + \frac{\beta_i^4}{\varkappa_i^4}\right]^{1/2} + 1\right]^{1/2}, \\ \varkappa_i &= \sqrt{T_i/2EI_i}\,, \quad \beta_i^4 = \mu_i \Omega^2/EI_i\,. \end{aligned}\right\} \qquad (6)$$

Die Lösungsanteile mit den (komplexen) Amplituden W_i^+ und W_i^- bezeichnen dabei rechts- bzw. linkslaufende dispersive Wellen, während die Anteile mit V_i^+ und V_i^- auf- oder abklingende Exponentialfunktionen darstellen.

Am Einbauort des Abstandhalters, vgl. Bild 4, gelten die Übergangsbedingungen

$$w_1(x, t)|_{x=0} = w_2(x, t)|_{x=0}\,,$$
$$\frac{\partial}{\partial x} w_1(x, t)|_{x=0} = \frac{\partial}{\partial x} w_2(x, t)|_{x=0}\,, \qquad (7a, b)$$

$$w_3(x, t)|_{x=0} = w_4(x, t)|_{x=0}\,,$$
$$\frac{\partial}{\partial x} w_3(x, t)|_{x=0} = \frac{\partial}{\partial x} w_4(x, t)|_{x=0}\,, \qquad (7c, d)$$

$$\frac{\partial^2}{\partial x^2} w_1(x,t)|_{x=0} - \frac{\partial^2}{\partial x^2} w_2(x,t)|_{x=0} = \frac{1}{EI_1} F_3(t), \tag{7e}$$

$$\frac{\partial^2}{\partial x^2} w_3(x,t)|_{x=0} - \frac{\partial^2}{\partial x^2} w_4(x,t)|_{x=0} = \frac{1}{EI_3} F_4(t), \tag{7f}$$

$$\frac{\partial^3}{\partial x^3} w_1(x,t)|_{x=0} - \frac{\partial^3}{\partial x^3} w_2(x,t)|_{x=0} = \frac{-1}{EI_1} F_1(t), \tag{7g}$$

$$\frac{\partial^3}{\partial x^3} w_3(x,t)|_{x=0} - \frac{\partial^3}{\partial x^3} w_4(x,t)|_{x=0} = \frac{-1}{EI_3} F_2(t), \tag{7h}$$

wobei für die Abschnitte 1 und 2 sowie 3 und 4 jeweils gleiche Systemparameter vorausgesetzt wurden.

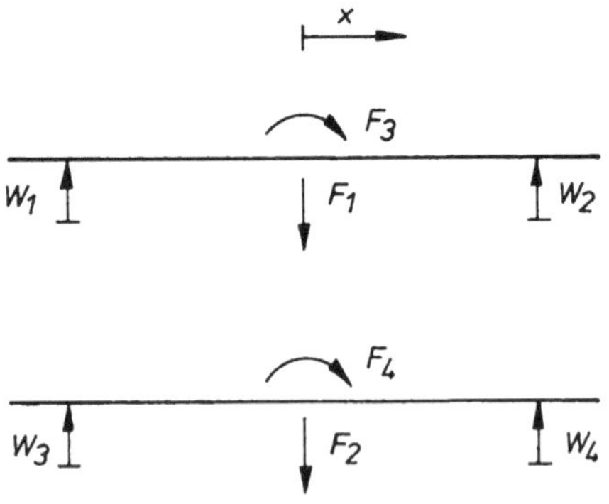

Bild 4

In den Lösungsdarstellungen (5) sind die Koeffizienten V_1^-, V_2^+, V_3^-, V_4^+ jeweils zu Null zu setzen, da sie zu unbeschränkt anwachsenden Lösungen führen. Die Koeffizienten W_1^+, W_2^-, W_3^+, W_4^- beschreiben die aus den Spannfeldern zum Abstandhalter hin einlaufenden Wellen. Sie werden im folgenden als gegeben betrachtet. Wir ordnen sie im Vektor

$$\boldsymbol{W}_E = [W_1^+, W_2^-, W_3^+, W_4^-]^T \tag{8}$$

an. Die restlichen Koeffizienten

$$\boldsymbol{W}_A = [W_1^-, V_1^+, W_2^+, V_2^-, W_3^-, V_3^+, W_4^+, V_4^-]^T \tag{9}$$

können aus (7) berechnet werden, indem man die Kraftamplituden $\hat{\boldsymbol{F}}$ mit Hilfe von (2) eliminiert. Dann folgt

$$\boldsymbol{A}(\Omega)\, \boldsymbol{W}_A = \boldsymbol{B}(\Omega)\, \boldsymbol{W}_E \tag{10}$$

mit den (komplexen) Matrizen $\boldsymbol{A}$ und $\boldsymbol{B}$. Für jede gegebene Kombination einlaufender Wellen kann jetzt die Seilschwingung berechnet werden. Damit sind dann auch die reflektierten Wellen und die sogenannten Nahfeldanteile bestimmt.

4. Die mittlere Dissipationsleistung

Die momentane Dissipationsleistung errechnet sich aus

$$P(t) = \sum_{i=1}^{4} F_i(t)\, \dot{q}_i(t) = \sum_{i=1}^{4} \operatorname{Re}[\hat{F}_i\, e^{j\Omega t}] \operatorname{Re}[\hat{\dot{q}}_i\, e^{j\Omega t}] \tag{11}$$

und die über eine Periode gemittelte Dissipationsleistung ergibt sich zu

$$\bar{P} = \frac{2\pi}{\Omega} \int_0^{2\pi/\Omega} P(t)\, dt = -\frac{1}{2} \Omega^2 \operatorname{Re}[\boldsymbol{W}^* \boldsymbol{Z}^*(\Omega)\, \boldsymbol{W}], \tag{12}$$

wobei

$$\boldsymbol{W} = \begin{bmatrix} W_1^+ + W_1^- + V_1^+ \\ W_3^+ + W_3^- + V_3^+ \\ -jk_1 W_1^+ + jk_1 W_1^- + l_1 V_1^+ \\ -jk_2 W_3^+ + jk_2 W_3^- + l_2 V_3^+ \end{bmatrix} \tag{13}$$

ist. Für $\boldsymbol{Z}(\Omega) = \boldsymbol{Z}^T(\Omega)$ folgt dann

$$\bar{P} = -\frac{1}{2} \Omega^2 [\boldsymbol{W}_{re}^T \boldsymbol{Z}_{re} \boldsymbol{W}_{re} + \boldsymbol{W}_{im}^T \boldsymbol{Z}_{im} \boldsymbol{W}_{im}]$$

$$= -\frac{1}{2} \Omega^2 [\boldsymbol{W}_{re}^T, \boldsymbol{W}_{im}^T] \left[\begin{array}{c|c} \boldsymbol{Z}_{re} & \boldsymbol{0} \\ \hline \boldsymbol{0} & \boldsymbol{Z}_{re} \end{array}\right] \left[\begin{array}{c} \boldsymbol{W}_{re} \\ \hline \boldsymbol{W}_{im} \end{array}\right], \tag{14}$$

wobei zur Abkürzung $\boldsymbol{W}_{re} = \operatorname{Re}[\boldsymbol{W}]$, usw. geschrieben wurde. Mit Hilfe der Projektionsmatrizen $\boldsymbol{C}_E$ und $\boldsymbol{C}_A$, die gemäß

$$\boldsymbol{W} = \boldsymbol{C}_E \boldsymbol{W}_E + \boldsymbol{C}_A \boldsymbol{W}_A \tag{15}$$

definiert sind, kann (14) als reelle quadratische Form

$$P = -\tfrac{1}{2} \Omega^2 \boldsymbol{u}^T \boldsymbol{Q} \boldsymbol{u} \tag{16}$$

mit

$$\boldsymbol{u} = [\operatorname{Re}[\boldsymbol{W}_E^T], \operatorname{Im}[\boldsymbol{W}_E^T]]^T \tag{17}$$

geschrieben werden. Dabei ist die Matrix $\boldsymbol{Q}$ durch

$$\boldsymbol{Q} = \boldsymbol{E}^T \left[\begin{array}{c|c} \boldsymbol{Z}_{re} & \boldsymbol{0} \\ \hline \boldsymbol{0} & \boldsymbol{Z}_{re} \end{array}\right] \boldsymbol{E} \tag{18}$$

mit

$$\boldsymbol{E} = \left[\begin{array}{c|c} \operatorname{Re}[\boldsymbol{C}_E + \boldsymbol{C}_A \boldsymbol{A}^{-1}(\Omega)\, \boldsymbol{B}(\Omega)] & -\operatorname{Im}[\boldsymbol{C}_E + \boldsymbol{C}_A \boldsymbol{A}^{-1}(\Omega)\, \boldsymbol{B}(\Omega)] \\ \hline \operatorname{Im}[\boldsymbol{C}_E + \boldsymbol{C}_A \boldsymbol{A}^{-1}(\Omega)\, \boldsymbol{B}(\Omega)] & \operatorname{Re}[\boldsymbol{C}_E + \boldsymbol{C}_A \boldsymbol{A}^{-1}(\Omega)\, \boldsymbol{B}(\Omega)] \end{array}\right] \tag{19}$$

gegeben. Für jede Kombination einlaufender Wellen kann jetzt die mittlere Dissipationsleistung $\bar{P}$ in Abhängigkeit der Systemparameter EI_i, μ_i, T_i sowie der Impedanzmatrix $\boldsymbol{Z}(\Omega)$ bestimmt werden.

5. Wahl optimaler Systemparameter

Die mittlere Dissipationsleistung hängt von den einlaufenden Wellen ab, so daß wir als Optimierungskriterium formulieren: „Die Systemparameter sind so zu wählen, daß für die ungünstigste Kombination einlaufender Wellen (gleicher Norm) der Betrag der mittleren Dissipationsleistung maximiert wird", also

$$\min_{\boldsymbol{u}^T \boldsymbol{u} = 1} [\boldsymbol{u}^T \boldsymbol{Q} \boldsymbol{u}] \to \text{Max!} \tag{20}$$

Das Minimum der quadratischen Form $\boldsymbol{u}^T \boldsymbol{Q} \boldsymbol{u}$ wird aber für $\boldsymbol{u}^T \boldsymbol{u} = 1$ genau dann angenommen, wenn $\boldsymbol{u}$ dem Eigenvektor von $\boldsymbol{Q}$ mit dem niedrigsten Eigenwert entspricht, so daß die Optimierung auch anhand der Vorschrift „Maximiere den kleinsten Eigenwert der Dissipationsmatrix $\boldsymbol{Q}$" vorgenommen werden kann.

Dieses Ergebnis läßt sich ohne weiteres auch auf Abstandhalter für andere Bündelleiter, z. B. Dreier- oder Viererbündel, übertragen, denn es ist stets möglich, die Energiedissipation als quadratische Form entsprechend (16) darzustellen.

6. Zusammenfassung

Es wurde ein Kriterium formuliert, mit dem selbstdämpfende Feldabstandhalter für Bündelleiter elektrischer Freileitungen optimiert werden können. Dabei sind die Systemparameter des Abstandhalters so zu wählen, daß der kleinste Eigenwert einer (reellen) symmetrischen Matrix maximiert wird. Damit kann gewährleistet werden, daß der Feldabstandhalter auch bei ungünstigen Kombinationen einlaufender Wellen einen genügend großen Energieanteil dissipiert. Das Optimierungskriterium ist unabhängig vom Einbauort und von den Randbedingungen am Spannfeldrand.

Literatur

1 DOWELL, E. H.: A modern course in aeroelasticity. Alphen aan den Rijn, Niederlande: Sijthoff & Noordhoff 1978.
2 EPRI (ed): Transmission line reference book, wind-induced conductor motion. Palo Alto, USA: Electric Power Research Institute 1979.
3 MÖCKS, L.; SCHMIDT, J.: Bemessung und Anordnung von selbstdämpfenden Feldabstandhaltern in Bündelleitungen. Elektrizitätswirtschaft **87** (1988), 1044–1048.
4 MÖCKS, L.: Bildung von Schwingungsinseln in Bündelleitern. Elektrizitätswirtschaft **69** (1970), 709–710.
5 HAGEDORN, P.: Ein einfaches Rechenmodell zur Berechnung winderregter Schwingungen an Hochspannungsfreileitungen mit Dämpfern. Ingenieur-Archiv **49** (1980), 161–177.

Eingegangen am 24. Mai 1989

Anschrift: Dr.-Ing. JÖRG WALLASCHEK, Technische Hochschule Darmstadt, Institut für Mechanik, Hochschulstraße 1, W-6100 Darmstadt, BRD

BOOK REVIEW

Böttcher, A; Silbermann, B., Analysis of Toeplitz Operators. Berlin, Akademie-Verlag 1989. 512 pp., DM 96,–. ISBN 3-05-500478-7

Das vorliegende umfangreiche Werk über Toeplitz-Operatoren ist von den Fachleuten sicherlich mit großer Spannung erwartet und nun mit großer Freude begrüßt worden. Mit sehr sorgfältiger Recherche ist die etwa vor 1988 erschienene Literatur berücksichtigt worden und der Stand der Theorie hier in sehr überzeugender Weise, ausgehend von den unerläßlichen funktionalanalytischen Grundlagen, dargelegt. Die Autoren sind hervorragend kompetente Fachleute, waren sie doch neben ihren vielen einschlägigen Originalarbeiten, etwa der letzten zehn Jahre, durch ihre Forschungsmonografie im Akademie-Verlag, Berlin, „Invertibility and Asymptotics of Toeplitz-Matrices“ schon weit bekannt geworden.

Die Toeplitz-Operatoren und die eng mit ihnen zusammenhängenden Wiener-Hopf-Operatoren haben eine lange Geschichte. Ihre Probleme sind häufig aus Anwendungsgebieten der Mathematik, wie der Strahlungs- und Beugungstheorie, der Signal- und Systemtheorie, aber auch der Informations- und Kontrolltheorie hervorgegangen. Sie haben aber auch innermathematisch viel Interesse gefunden und im Zusammenhang genauerer Aussagen über die Lösungstheorie spezieller Operatorgleichungen mit Differenzen- und Summenkernfunktionen der Integraloperatoren und deren diskrete endlich-dimensionale Approximationen in den Toeplitz- und Hankelmatrizen ein weites Feld gefunden und immer wieder neue subtile Fragestellungen ermöglicht. Das vorliegende ca. 500 Seiten umfassende Werk gibt einen tiefen Einblick in die rein mathematische Theorie, verweilt aber auch längere Zeit bei den Approximationsverfahren, speziell im siebten Kapitel über die „Methode der endlichen Abschnitte“, die aber auch später in komplizierteren Situationen, wie z. B. beim „Viertelebenenproblem“, im Kapitel 8 oder bei der sehr spezifizierten Asymptotik der Toeplitz-Determinanten im Kapitel 10 erscheint.

Es war sicherlich sehr schwierig für die Autoren, eine Stoffabgrenzung vorzunehmen. Sie haben m. E. sehr sorgfältig darüber nachgedacht und eine gute Lösung gefunden, die freilich eindeutig die Theorie der Toeplitz-Operatoren bevorzugt. Hätten sie Anwendungsgebiete und echte numerische Aspekte einbezogen, hätte dies wohl einen zweiten Band erfordert. Der Leser wünschte sich vielleicht dennoch einige wenige Abschnitte über wichtige Anwendungsgebiete mit klassischen Fragestellungen, um vielleicht einen noch größeren Leserkreis für dieses so spannende, lebendige Gebiet der Operatortheorie zu gewinnen.

Im folgenden soll ein wenig auf den Inhalt der zehn Kapitel eingegangen werden: Im ersten Kapitel sind die funktionalanalytischen Fakten über die s-Zahlen, die Schatten-von-Neumann-Klassen, über Operatordeterminanten, Fredholmoperatoren, die Indexformel insbesondere in der Beziehung von Systemen von Gleichungen mit Operatoren aus verschiedenen Banachalgebren zusammengestellt, die üblicherweise nicht in einem einführenden Lehrbuch stehen.

Hier sind auch die für die spätere Analyse so wichtigen „lokalen Prinzipien“ von Gohberg-Krupnik und Allan-Douglas aufgenommen worden und die grundlegenden Sätze der L^p- und Hardyraum (H^p-)-Theorie mit der harmonischen Fortsetzung von Einheitskreisperipherie Π in die Kreisscheibe $\mathbb{D}$, der Innen-, Außen-Faktorisierung aufgenommen worden. Der Leser findet auch die Hunt-Muckenhoupt-Wheeden-Bedingung für gewichtete L^p-Räume und die Klassen BMO und VMO von Funktionen mit beschränkter mittlerer Schwankung. In den Bemerkungen am Ende jedes Kapitels finden sich sehr sorgfältig nachgewiesene historische Bemerkungen, so zum ersten Kapitel zur Geschichte des lokalen Prinzips auf den Seiten 45/46.

Mit diesem Rüstzeug kann sich der Leser im zweiten Kapitel mit der grundlegenden Theorie vertraut machen: Definition der Multiplikations-, Toeplitz- und Hankeloperatoren mit den wichtigen Sätzen von Brown-Halmos und Nehari, die die Invertierbarkeit in Abhängigkeit von elementaren Symboleigenschaften betreffen. Die wichtigen spektralen Einschließungssätze von Hartmann-Wintner und Brown-Halmos findet man ebenso wie den von Coburn gefundenen Zusammenhang zwischen Fredholmeigenschaft und Invertierbarkeit von Toeplitz-Operatoren für L^∞-Symbole in l^p und $H^p, 1 < p < \infty$. Dann wird die Kompaktheit der Hankeloperatoren mit Symbolen aus dem „Douglasraum“ C und H^∞ gezeigt. Vor der Einführung von matrixwertigen Symbolen im letzten Abschnitt findet der Leser eine sorgfältige, detaillierte Einführung der lokalen Methoden für skalare Toeplitz-Operatoren. Es wird hier gezeigt, daß ein T-Operator mit „lokal sektoriellem Symbol“ über H^2 ein Fredholm-Operator ist.

Im Kapitel 3 steht die Symbolanalysis im Vordergrund. Es werden zunächst die Begriffe „analytisch und geometrisch sektorieller Symbolmatrizen“ eingeführt und diese charakterisiert. Viele spätere verfeinerte Untersuchungen benötigen die „asymptotische Multiplikativität“, die sich weitgehend auf die „Approximation der Identität (Eins)“ stützt. Dann werden die „quasi-stetigen Funktionen“, QC, auf den Kreis Π eingeführt. Mittels einer harmonischen Approximation über das Poisson-Integral kann eine Algebraisierung, „Verwesentlichung“ und Lokalisation von Subalgebren zu den Symbolunteralgebren von $L^\infty_{N\times N}(\Pi)$ vorgenommen werden. Die lokale Sektorialität kann dann weitgehend charakterisiert werden.

Das Kapitel 4 ist den Toeplitz-Operatoren im Hardy-Raum H^2 gewidmet, während die Autoren die analogen bekannten Sachverhalte in den H^p-Räumen ($p \neq 2$) im darauffolgenden Kapitel 5 abhandeln. Nach Klärung der Fredholmeigenschaft im Satz 4.12 wird dann die Frage nach der „stabilen Konvergenz“ approximierender Operatorscharen und daran anschließend die Indexberechnung erläutert. Die „transfinite Zerlegung“ des Raumes $M(L^\infty)$ der maximalen Ideale in $L^\infty(\Pi)$ gestattet dann, lokale Toeplitz-Operatoren und spezielle Symbolklassen derselben zu untersuchen. Der Satz 4.48 (von Axler) ist hier ein sehr wichtiges Resultat. Schließlich sind dann noch die Gelfandtopologien auf den Schilowrändern einiger Symbolräume beschrieben.

Im fünften Kapitel (H^p-Theorie) sind besonders erwähnenswert einerseits das Resultat 5.5 (von Simonenko) über die Fredholmeigenschaft eines Toeplitz-Operators und der verallgemeinerten Faktorisierbarkeit seines Symbols $a \in GL^\infty$ und andererseits in den „Chwedelidse-gewichteten L^p-Räumen“ die „lokal p,ϱ-sektoriellen Symbole“: Ist das Symbol a sogar aus dem Douglasraum, dann ist der T-Operator fredholmsch auf $L^p(\varrho)$, wenn sein Symbol a über der Douglasklasse p,ϱ-sektoriell ist. Die (stärkere) Invertierbarkeit von

$T(a)$ über H^p (ohne Gewicht) wird durch den Satz 5.22 (von Krupnik) garantiert. Die Lokalisation bei stückweise stetigen Symbolen a führt dann in diesen L^p-Räumen zu den bekannten – von Gohberg-Krupnik eingeführten – $a_{p,\varrho}(t, \mu)$-Symbolen. Hier gibt es noch eine Reihe offener Probleme (s. S. 243!). Eine genaue Kenntnis der Toeplitz-Operatoren hat man bei sog. „Fisher-Hartwig-Symbolen".

Das „Folgenanalogon" der Toeplitzoperatoren wird auf l^p im Kapitel 6 studiert. Zunächst werden die wichtigsten Eigenschaften der l^p-Multiplikatoren aufgelistet, ehe die Autoren die folgenden Symbolklassen betrachten: stetige, stückweise stetige und analytische Symbole. Hier sind bereits viele, schwer im Detail zu beschreibende, Einzelresultate bekannt. (Vinogradov, Verbitski, Pomp, Horbach).

Im Kapitel 7 greifen die Autoren das endliche Approximationsproblem an, indem sie erst einmal generell Projektionsverfahren für Operatoren zwischen Banachräumen erklären, diese dann als „endliche Abschnittsmethode" für Toeplitz-Operatoren mit unendlichen Matrizen und die Hilfsmittel der Algebraisierung und Verwesentlichung einführen. Die Klassen der approximierbaren Operatoren $T(a) + K$, K: kompakt, werden dann für $C + H^\infty$-Symbole, für lokal sektorielle und für stückweise stetige Symbole – getrennt nach l^p- und H^p-Theorie – untersucht. Es resultieren hier sehr schöne charakterisierende Sätze von Krupnik und Verbitski. Für Fisher-Hartwig-Symbole steht ebenfalls eine verfeinerte Theorie zur Verfügung. Bemerkenswert ist das Ergebnis von Treil im letzten Abschnitt, daß es invertierbare Toeplitz-Operatoren auf dem Hardyraum H^2 mit L^∞-Symbol gibt, für die die endliche Abschnittsmethode versagt (Theorem 7.92).

Im Kapitel 8 erweitern die Autoren nun ihre Fragestellungen auf den Fall einer Viertelebene bzw. des Torus $\Pi^2 := \Pi \times \Pi$. Zuerst werden dort Funktionenklassen und Tensorprodukte zu Räumen auf Π und zugehörige Operatoren zwischen solchen Funktionenräumen eingeführt. Dann können einfache Eigenschaften von zweidimensionalen Toeplitz-Operaten, $T_2(a)$, zwischen den verallgemeinerten Hardyräumen $H^p_N(\Pi^2)$ mit $T_2(a)$ definiert durch $\varphi \mapsto P_{++}(a\varphi)$ bewiesen werden. Hierbei spielen auch die „Halbraum-Toeplitz-Operatoren" eine wichtige Rolle, da sie später zur Konstruktion von Fredholmregularisatoren benutzt werden. Zuerst werden dann stetige Symbole zugrunde gelegt, ehe das allgemeine Invertierbarkeitsproblem angeschnitten wird. Dabei sind bisher nur für sehr spezielle Symbolklassen definitive Resultate erzielt worden (u. a. von Douglas-Howe, Böttcher, Simonenko). Die „bilokale" Fredholmtheorie erlaubt mit einem „bilokalen Lokalisationsprinzip" weitergehende Aussagen, wie sie z. B. über die Fredholmeigenschaft in den Sätzen 8.43 und 8.47 enthalten sind. Die endliche Abschnitte-Methode ist nach der Kozak- wie auch der bilokalen Theorie hier dargelegt. Der Schlußabschnitt enthält die verallgemeinerten Formulierungen für den höherdimensionalen Fall Π^k.

Das Kapitel 9 enthält die entsprechende Theorie der Wiener-Hopf-Integraloperatoren auf der Halbachse und dem Quadranten. Die klassische Krein-Gohberg-Theorie in $L^p(\mathbb{R}_+)$-Räumen ist zunächst für Symbole der Wiener-Algebra $W(\mathbb{R})$ dargestellt, dann werden auch stückweise stetige und oszillierende Symbole zugelassen, wie sie z. B. bei Integrodifferenzengleichungen auftreten. Literatur zu den stückweise stetigen fastperiodischen Symbolen auf der Achse ist angegeben: Hier ist ein wichtiges Anwendungsgebiet, u. a. in der mathematischen Beugungstheorie, zu sehen. Die Methode der endlichen Abschnitte hier beruht auf Scharen von Faltungsgleichungen in wachsenden endlichen Intervallen $(0, \tau)$. Satz 9.42 charakterisiert dann die Approximierbarkeit von invertierbaren Wiener-Hopf-Gleichungen auf $L^p(\mathbb{R}_+)$ für stückweise stetige Symbole. Hier sind noch einige Fragen bzgl. des Spektrums solcher (singulärer) Faltungsoperatoren im L^p-Fall offen! Bei den Viertelebenen-Faltungsoperatoren gibt es bereits Fredholmresultate (u. a. von Dudučava, Böttcher und dem Referenten), aber das allgemeine Invertierbarkeitsproblem ist immer noch offen.

Das Schlußkapitel 10 enthält nun viele Einzelheiten über Toeplitz-Determinanten, der auf Szegö (1915) i. W. zurückgehenden Theorie über $D_n(A) := \det (a_{j-k})_{j,k=0,\dots,n}$; $n \in \mathbb{N}$; der $(n + 1)$-ter Ordnung Abschnittsdeterminanten einer unendlichen Toeplitz-Matrix $(a_{j-k})_{j,k\in\mathbb{N}_0}$. Der erste Szegöische Grenzwertsatz 10.6 sagt aus, daß der Grenzwert $\lim_{n\to\infty} D_n(a)/D_{n-1}(a) = G(a) = 0$ existiert für Symbolfunktionen a aus der Douglas-Klasse $(C + H^\infty)_{N\times N}$ oder für lokal sektorielle Matrixsymbole über den quasistetigen Funktionen, wenn $T(a)$ und $T(\tilde{a})$ invertierbar über H^2_N sind (mit $\tilde{a}(t) := a(t^{-1})$ $t \in \Pi$). Alle weiteren Untersuchungen und Ergebnisse in diesem Kapitel zielen auf die Angabe von Bedingungen an die Symbolfunktionen a, die Eigenschaften der Toeplitz-Operatoren auf geeigneten Banachräumen und auf genauere Aussagen über die Rate der Konvergenz der Quotienten $D_n(a)/D_{n-1}(a)$ gegen den Grenzwert. Hier ist Szegö's starker Grenzwertsatz 10.29 in den Kreinschen Symbolklassen wichtig. Eine höhere Asymptotik für $\log D_n(a)$ wurde u. a. von den Autoren des vorliegenden Buches erzielt (Satz 10.35). Der Zusammenhang zwischen Faktorisierbarkeit der Matrixsymbole in verschiedenen Klassen ist in wichtigen Fällen diskutiert, wenn der Operatorindex nicht verschwindet. Für Fischer-Hartwig und unbeschränkte lokal sektorielle Symbole sind einige Ergebnisse nun bekannt. Am Ende des Werkes sind noch die Zusammenhänge zwischen der asymptotischen Verteilung der Eigenwerte mit den Spuren von Toeplitz-Matrizen sowie der höherdimensionale Fall angesprochen.

Das Analogon im kontinuierlichen Wiener-Hopf-Fall wird von der Achieser-Kac-Formel beherrscht. Die zugehörigen „Wiener-Hopf-Determinanten" bei Fisher-Hartwig-Symbolen und Verallgemeinerung der Formel auf höhere Dimensionen stellen neben ausführlichen Kommentaren und Literaturnachweisen, hilfreicher Symbol-, Namens- und Gegenstandsverzeichnissen den Schluß des Buches dar.

Das Buch ist sehr sorgfältig verfaßt und gedruckt worden und stellt m. E. eine Perle in der modernen banachalgebraischen Operatortheorie dar. Die Lektüre ist an jeder Stelle ein Genuß, auch wenn vom Leser einiges abgefordert wird. Die meisterliche Beherrschung der Symboltechnik und das sichere Umgehen mit den involvierten Abschätzungen weist die Autoren neben ihrem immensen Überblick und Ideenreichtum, der sich in vielen eigenen wichtigen Beiträgen zu diesen reizvollem Thema zeigt, auch als „sehr geschickte Handwerker" aus. Ich wünsche dem Werk eine rasche und weite Verbreitung.

Darmstadt E. MEISTER

NEW BOOKS

Riccardi, L. M. (ed.), Lectures in Applied Mathematics and Informatics. Manchester/New York, Manchester University Press 1990. XI, 362 pp., Ł 70.00. ISBN 0-7190-2671-7

Bochmann, D.; Ubar, R. (Hrsg.), Fehler in Automaten. Berlin, Verlag Technik 1989. 216 S., 96 Bilder, 17 Taf., DM 30,–. ISBN 3-341-00683-4

Dautray, R.; Lions, J.-L., Mathematical Analysis and Numerical Methods for Science and Technology. Vol. 3: Spectral Theory and Applications. Berlin etc., Springer-Verlag 1990. X, 515 pp., DM 198,00. ISBN 3-540-50208-4

Claßen, L.; Wiesner, U., Wissensspeicher 80286-Programmierung. Berlin, Verlag Technik 1990. 240 S., DM 29,80. ISBN 3-341-00867-5 (Technische Informatik)

vom Scheidt, J., Stochastic Equations of Mathematical Physics. Berlin, Akademie-Verlag 1990. 475 pp., 147 figs., 33 tables, DM 90.–. ISBN 3-05-500609-7 (Math. Lehrbücher und Monographien. II. Abt.: Math. Monographien 76)

Wiggins, S., Introduction to Applied Nonlinear Dynamical Systems and Chaos. Berlin etc., Springer-Verlag 1990. 672 pp., 291 figs., DM 98,00. ISBN 3-540-97003-7 (Texts in Applied Mathematics 2)

Gericke, H., Mathematik im Abendland. Von den römischen Feldmessern bis zu Descartes. Berlin etc., Springer-Verlag 1990. XIII, 352 S., 143 Abb., DM 128,00. ISBN 3-540-51206-3

Arrowsmith, D. K.; Place, C. M., An Introduction to Dynamical Systems. Cambridge etc., Cambridge University Press 1990. 423 pp., Ł 15.00. ISBN 0-521-31650-2

Smith, P.; Smith, R. C., Mechanics. 2nd edition. Chichester etc., John Wiley & Sons 1990. XIX, 321 pp., Ł 13.95. ISBN 0-471-92737-6

Szczepiński, W. (ed.), Experimental Methods in Mechanics of Solids. Amsterdam etc., Elsevier and Warszawa, PWN-Polish Scientific Publishers 1990. XVII, 705 pp., Dfl. 420.00. ISBN 0-444-98883-1 (Materials Science Monographs 51)

Brebbia, C. A.; Connor, J. J. (eds.), Advances in Boundary Elements. Vol. 2: Field and Fluid Flow Solutions. Berlin etc., Springer-Verlag 1989. IX, 484 pp., DM 198,00. ISBN 3-540-51449-X

Olver, P. J.; Sattinger, D. H. (eds.), Solitons in Physics, Mathematics and Nonlinear Optics. Berlin etc., Springer-Verlag 1990. XIII, 215 pp., 37 figs., DM 58,00. ISBN 3-540-97309-5 (The IMA Volumes in Mathematics and its Applications 25)

Lützen, J., Joseph Liouville 1809—1882. Master of Pure and Applied Mathematics. Berlin etc., Springer-Verlag 1990. XIX, 884 pp., 96 figs., DM 228,00. ISBN 3-540-97180-7 (Studies in the History of Mathematics and Physical Sciences 15)

Szemplińska-Stupnicka, W., The Behavior of Nonlinear Vibrating Systems. I: Fundamental Concepts and Methods: Applications to Single-Degree-of-Freedom Systems. II: Advanced Concepts and Applications to Multi-Degree-of-Freedom Systems. Dordrecht etc., Kluwer Academic Publishers 1990. XVII, 253 pp., Dfl. 170.00; XVIII, 330 pp., Dfl. 170.00. Set ISBN 0-7923-0370-9 (Mechanics: Dynamical Systems 12 and 13)

Stephani, H., Differential Equations. Their Solution Using Symmetries. Cambridge etc., Cambridge University Press 1989. XII, 260 pp., Ł 12.95 P/b. ISBN 0-521-36689-5

Broer, H. W.; Verhulst, F., Dynamische Systemen en Chaos, een revolutie vanuit de wiskunde. Utrecht, Epsilon Uitgaven 1990. XIV, 349 pp., Dfl. 44.50. ISBN 90-5041-020-0

Le Veque, R. J., Numerical Methods for Conservation Laws. Basel etc., Birkhäuser Verlag 1990. 214 pp., sfr 38.00. ISBN 3-7643-2464-3

Rotta, J. C., Die Aerodynamische Versuchsanstalt in Göttingen, ein Werk Ludwig Prandtls. Ihre Geschichte von den Anfängen bis 1925. Göttingen, Vandenhoeck & Ruprecht 1990. XII, 332 S., 233 Abb., 5 Tab., DM 88,—. ISBN 3-525-42808-1

Schatz, A. H.; Thomée, V.; Wendland, W. L., Mathematical Theory of Finite and Boundary Element Methods. Basel etc., Birkhäuser Verlag 1990. 276 pp., sfr 52.00. ISBN 3-7643-2211-X (DMV Seminar 15)

Gürlebeck, K.; Sprößig, W., Quaternionic Analysis and Elliptic Boundary Value Problems. Basel etc., Birkhäuser Verlag 1990. 253 pp., sfr 78.00. ISBN 3-7643-2382-5 (ISNM 89)

Donley, M. G.; Spanos, P. D., Dynamic Analysis of Non-Linear Structures by the Method of Statistical Quadratization. Berlin etc., Springer-Verlag 1990. VII, 172 pp., 90 figs., 4 tab., DM 50,00. ISBN 3-540-52743-5 (Lecture Notes in Engineering 57)

Frank, L. S., Singular Perturbations. I: Spaces and Singular Perturbations on Manifolds without Boundary. Amsterdam etc., North-Holland 1990. XXIV, 556 pp., Dfl. 175.00. ISBN 0-444-88134-4 (Studies in Mathematics and its Applications 23)

Schipp, F.; Wade, W. R.; Simon, P., Walsh Series. An Introduction to Dyadic Harmonic Analysis. Budapest, Akadémiai Kiadó 1990. X, 560 pp., Ft 1120,—. ISBN 963-05-5880-7

Rutishauser, H., Lectures on Numerical Mathematics. Boston etc., Birkhäuser 1990. 546 pp., sfr 88.—. ISBN 3-7643-3491-6

Lam, L.; Morris, H. C. (eds.), Nonlinear Structures in Physical Systems. Pattern Formation, Chaos, and Waves. Proceedings of the Second Woodward Conference, San Jose State University, November 17—18, 1989. Berlin etc., Springer-Verlag 1990. X, 331 pp., 135 figs., DM 98,00. ISBN 3-540-97344-3

Banichuk, N. V., Introduction to Optimization of Structures. Berlin etc., Springer-Verlag 1990. X, 300 pp., 66 figs., DM 178,00. ISBN 3-540-97212-9

Frommer, A., Lösung linearer Gleichungssysteme auf Parallelrechnern. Braunschweig, Friedr. Vieweg & Sohn 1990. 267 S. ISBN 3-528-06397-1

Degasperis, A.; Fordy, A. P.; Lakshmanan, M. (eds.), Nonlinear Evolution Equations: Integrability and Spectral Methods. Manchester/New York, Manchester University Press 1990. 619 pp., Ł 65.00. ISBN 0-7190-3273-3 (Proceedings in Nonlinear Science)

Arnol'd, V. I., Huygens and Barrow, Newton and Hooke. Pioneers in Mathematical Analysis and Catastrophe Theory from Evolvents to Quasicrystals. Basel etc., Birkhäuser Verlag 1990. 118 pp., sfr 24.00. ISBN 3-7643-2383-3

Casciati, F.; Elishakoff, I.; Roberts, J. B. (eds.), Nonlinear Structural Systems under Random Conditions. Proceedings of the European Mechanics Colloquium, Euromech 250, Como, Italy, June 19—23, 1989. Amsterdam etc., Elsevier 1990. XII, 386 pp., Dfl. 280.00. ISBN 0-444-88803-9

Dautray, R.; Lions, J.-L., Mathematical Analysis and Numerical Methods for Science and Technology. Vol. 4: Integral Equations and Numerical Methods. Berlin etc., Springer-Verlag 1990. X, 465 pp., 67 figs., DM 198,00. ISBN 3-540-50209-2

Porter, D.; Stirling, D. S., Integral Equations. A Practical Treatment, from Spectral Theory to Applications. Cambridge etc., Cambridge University Press 1990. XI, 372 pp., Ł 15.00 H/b. ISBN 0-521-33742-9 (Cambridge Texts in Applied Mathematics)

Bruce, J. W.; Giblin, P. J.; Rippon, P. J., Microcomputers and Mathematics. Cambridge etc., Cambridge University Press 1990. XVI, 425 pp., Ł 17.50 P/b. ISBN 0-521-31238-8

Kaashoek, J. F., Modeling one dimensional pattern formation by anti-diffusion. Amsterdam, Stichting Mathematisch Centrum 1990. 271 pp., Dfl. 67.00. ISBN 90-6196-389-3 (CWI Tract 72)

Freud, R. (Hgb.), Große Augenblicke aus der Geschichte der Mathematik. Budapest, Akadémiai Kiadó 1990. 263 S. ISBN 963-05-5640-5

Simpson, A., Das WordPerfect 5.1 Buch. Düsseldorf etc., SYBEX-Verlag 1990. XXIX, 993 S., DM 79,—. ISBN 3-88745-823-0

OBITUARY

Zum Tode von Lothar Collatz

Einer der bedeutenden Mathematiker dieses Jahrhunderts, Lothar Collatz, ist tot.

Am 26. September 1990, inmitten der Arbeit, während eines Kongresses in Varna/Bulgarien, verstarb er 80-jährig plötzlich und unerwartet.

Siechtum und geistige Müdigkeit blieben ihm erspart, verstand er es doch, sich mit Freude, aber auch mit großer Beharrlichkeit, durch Schwimmen und Skilaufen, durch Meidung schädlicher Genußmittel, durch Entspannung beim Skizzieren von Landschaften und bei Hobbymalerei die für seine wissenschaftliche Arbeit erforderliche geistige und körperliche Spannkraft zu erhalten.

Bis zuletzt hielt er Vorlesungen, wobei die Themen im Laufe der Jahre nahezu alle Bereiche der numerischen und der nichtnumerischen Angewandten Mathematik berührten. Seine Hörer dankten es ihm durch Treue und Anhänglichkeit.

Alljährlich veranstaltete Lothar Collatz im Zusammenwirken mit wechselnden Partnern angesehene Tagungen im Mathematischen Forschungsinstitut Oberwolfach, und umgekehrt besuchte er bis zuletzt zahlreiche internationale Tagungen und Kolloquien, in der Regel als geladener Hauptreferent.

Von dieser Reisetätigkeit zeugten auch seine Jahr für Jahr getreulich an seine Freunde und Kollegen in aller Welt versandten Weihnachtsgrüße, die aus selbstgefertigten Skizzen markanter Sehenswürdigkeiten der im jeweiligen Jahr besuchten Weltregionen bestanden. Weder politische noch klimatische Widrigkeiten konnten ihn aufhalten, wenn es galt, wissenschaftliche und dabei zugleich menschliche Kontakte zu knüpfen, sein Wissen weiterzugeben, zu lernen und junge Menschen anzuregen.

Scheinbar ein unpolitischer Mensch, der trotz seiner engen Kontakte zu Courant und anderen später verfolgten Mathematikern auch in nationalsozialistischer Zeit persönlich weitgehend unbehelligt blieb (wenn man davon absieht, daß er sich veranlaßt sah, seine Habilitationsschrift und eine Nachfolgearbeit in Bierberbachs schlimmer „Deutsche Mathematik" zu publizieren [3], [4]), war er dennoch im besten Sinne von eminenter politischer Wirksamkeit im Interesse internationaler Verständigung und Zusammenarbeit.

Die wissenschaftliche Welt dankte es ihm durch zahlreiche Auszeichnungen: Er war Träger von Verdienstmedaillen verschiedener osteuropäischer Universitäten sowie Ehrendoktor der Universität von São Paulo (1956), der TU Wien (1967), der Universität Dundee (1974), der Brunel University (1977), der TH Hannover (1981), der Universität Augsburg (1985) und der TU Dresden (1990). Er wurde zum Mitglied der Akademie der Naturforscher „Leopoldina" in Halle gewählt (1962) und war Ehrenmitglied der Akademien von Modena (1966) und Bologna (1978). Auch die 300-jährige Hamburger Mathematische Gesellschaft sowie die GAMM, deren Schriftführer er von 1958–1961 und deren Vorsitzender er von 1964–1967 gewesen war, erhoben ihn zum Ehrenmitglied.

In den letzten Jahren betonte Lothar Collatz anläßlich solcher Würdigungen mehrfach, daß Ehrungen im Grunde in Frage gestellt werden müßten, seien Begabungen doch durch Gott geschenkt, schöpferische Ideen also eher Gnade und Verpflichtung denn Verdienst. Aber natürlich entsprangen seine Leistungen nicht nur seiner mathematischen Begabung, dem Aufwachsen in einem wissenschaftlich geprägten Elternhaus, seinem hervorragenden Gedächtnis und einem seiner Liebenswürdigkeit nur selten entgegenstehenden starken Willen – gegebenenfalls auch Kampfbereitschaft – sondern sie waren auch Ergebnis seines immensen Fleißes und seiner Fähigkeit, neue Entwicklungen frühzeitig analysieren und somit in unbedeutende und weittragende scheiden zu können. Und er konnte fähige junge Menschen für die wichtigen Entwicklungen als begeisternder Lehrer und aktiv solche Entwicklungen mitbestimmender oder selbst bewirkender Forscher gewinnen.

Eine liebevolle Familie hielt häusliche Sorgen möglichst von ihm fern und schuf so die äußeren Voraussetzungen für die notwendige konstruktive Muße und Konzentration.

Zu den von Collatz initiierten wichtigen Entwicklungen in der Numerischen Mathematik zählen u. a. die ε-Schemata [2] (Vorläufer der Untersuchungen über Numerische Stabilität in der Numerik partieller Differentialgleichungen), die erste Befassung mit (periodischen) Splines (damals noch nicht so genannt) [5], die Mehrstellenverfahren (d. h. das Heranziehen der Differentialgleichung an mehreren Punkten bei der numerischen Lösung dieser Gleichung) ([6], S. 275ff), die Prägung des Begriffs der Operatoren monotoner Art und die darauf gründenden Einschließungsmethoden [7], die Einführung der H-Mengen in die Approximationstheorie [8] usw.

Das wissenschaftliche Gesamtwerk weist ca. 230 Veröffentlichungen aus. Folgt man der von seinem frühen Doktoranden, Günter Bertram, vorgenommenen Einteilung dieses Gesamtwerkes in Perioden [9] (vgl. auch [10]), so stehen zunächst die Differenzenverfahren und Eigenwertprobleme im Vordergrund, wobei sich wohl insbesondere der Einfluß von Courant und von v. Mises aus gemeinsamer Göttinger, bzw. Berliner Zeit auswirkt. Diese Periode reicht etwa von 1933–1951. Bereits hier wie auch später ließ es Collatz nie bei der alleinigen Vorlage richtungsweisender Originalarbeiten bewenden, sondern machte sich um seine Wissenschaft, deren Verbreitung, ihr Eindringen in Anwendungsgebiete und ihre Anerkennung in einer der Numerik damals zuweilen abwartend gegenüberstehenden Mathematiker-Welt durch seine in vielen Sprachen übersetzten, in mehreren Auflagen erschienen Monographien und Lehrbücher verdient.

Die erwähnte erste Periode brachte sein schon zitiertes Buch über „Eigenwertprobleme und ihre numerische Behandlung" (1945; 2. Auflage unter dem Titel „Eigenwertaufgaben mit technischen Anwendungen", 1949) hervor sowie die Bücher „Differentialgleichungen für Ingenieure" (1949) und „Numerische Behandlung von Differentialgleichungen" (1951).

Zur zweiten Periode zählt die Zeit von 1952–1966, die vornehmlich funktionalanalytischen Hintergrund hatte (Monotoniesätze, Iteration, Fixpunktsätze, Optimierung) und in dem bedeutenden Buch über „Funktionalanalysis und Numerische Mathematik" (1964) gipfelte. Das mit seinem Schüler W. Wetterling verfaßte Buch über „Optimierungsaufgaben" (1966) schließt diese Periode ab.

Eine dritte Periode, etwa 1965–1973, ist vor allem der Approximationstheorie gewidmet und schlägt sich in dem gleichnamigen Buch (1973) nieder, das mit W. Krabs – also ebenfalls mit einem seiner Schüler – geschrieben wurde.

Die Nutzbarmachung der Graphentheorie für numerische Zwecke und die verstärkte Hinwendung zu seiner alten mathematischen Liebhaberei [1], der Klassifizierung geometrischer Ornamente, kann als vierte und letzte Periode angesprochen werden.

Indessen muß jeder derartige Versuch einer Einteilung in Schaffensperioden angesichts der großen Spannweite dieses Forschers unvollkommen bleiben, gibt es doch auch zahlreiche Arbeiten, z. B. zum mathematischen Unterricht, zu Strömungsproblemen, zur instrumentellen Mathematik, zu Verzweigungsproblemen, zur Methode der Finiten Elemente usw., die außerhalb dieser Schwerpunkte stehen oder zeitlich weit außerhalb dieser Perioden publiziert wurden.

Natürlich gehörte Collatz auch zu jenen Mathematikern, die sehr früh die Bedeutung der in den 50er Jahren aufkommenden Großrechner für die Wissenschaft – und beileibe nicht nur für die Numerik – erkannten, und in den eigenen, besonders aber auch in anderen Wissenschaftsbereichen für das neue Instrument warben. Auf diese Weise schuf er in Hamburg 1958 eine „Abteilung für elektronische Rechenanlagen" und damit das erste Hochschulrechenzentrum, das mit einer IBM 650, der damals wohl leistungsfähigsten Maschine, ausgestattet wurde (zeitgleich mit einer entsprechenden Ausstattung der Universität Hannover).

Sein mathematischer Geist arbeitete ruhelos: Stets führte er in den Jackentaschen kleine Bleistiftreste und etwas Papier mit sich, um Gedanken und Ideen jederzeit sofort festhalten und später ausarbeiten zu können, und die zukünftige Auswertung seines wissenschaftlichen Nachlasses und seiner in winziger Schrift in Taschenkalenderform niedergelegten Tagebuchaufzeichnungen stellt eine wichtige Aufgabe dar.

Und natürlich war er als Mitherausgeber oder als Angehöriger des Beratungsstabes zahlreicher wissenschaftlicher Zeitschriften wie als Berichterstatter für Referaten-Organe geschätzt und diente auch auf diese Weise der Einhaltung hohen Forschungsniveaus.

Zwar fanden Philosophie, Geschichte, Theologie, Literatur bei Lothar Collatz nur allgemeines Interesse und eine glänzende Rhetorik stand ihm nur in der Mathematik zu Gebote, doch eine Fülle von Geschichten über Begegnungen mit anderen Wissenschaftlern oder Berichte über vielerlei ernste und lustige Erlebnisse machten ihn auch außerhalb des Hörsaals oder des Kolloquiums meist bald zum Mittelpunkt.

Am 6. Juli 1910 in Arnsberg/Westfalen geboren, begann sein wissenschaftlicher Werdegang mit dem Studium der Mathematik und Physik in den Jahren 1928 bis 1933 an den Universitäten Greifswald (H. Kneser, Süss), München (Perron, Carathéodory), Göttingen (Hilbert, Courant) und Berlin (v. Mises, Schrödinger, Bieberbach, E. Schmidt), wo er 1933 die Lehramtsprüfung ablegte und 1935 mit einer Arbeit über „Differenzenverfahren mit höherer Approximation" zum Dr. phil. promovierte. Er nahm dann eine Assistentenstelle bei Pöschl in Karlsruhe an, wo er sich 1937 mit der schon erwähnten Schrift [3] über einen Konvergenzbeweis und Fehlerabschätzung für Differenzverfahren bei Eigenwertaufgaben gewöhnlicher Differentialgleichungen zweiter und vierter Ordnung habilitierte und 1938 zum Privatdozenten ernannt wurde. In Karlsruhe lernte er auch seine Frau Martha kennen. 1943 wurde Collatz als Ordentlicher Professor für Mathematik nach Hannover berufen.

An der Universität Hamburg hatte man seit der Universitätsgründung 1919 an dem Prinzip nur *einer* Mathematik und damit auch nur eines Instituts, des Mathematischen Seminars, fesgehalten, und dieses Institut wurde durch die ordentlichen Professoren der Mathematik kollegial geleitet. Die wachsende Bedeutung der Numerik führte zur Gründung einer „Arbeitsstelle für Praktische Mathematik" innerhalb des Mathematischen Seminars, und als Leiter dieser Arbeitsstelle und zugleich als weiterer Direktor des Seminars wurde Collatz 1952 auf eine ordentliche Professur für Mathematik berufen. Er nahm diesen Ruf an und blieb Hamburg trotz späterer weiterer Rufe bis zu seinem Tode, also für nahezu die Hälfte seines Lebens, treu. In der Überzeugung, seiner Aufgabe zur Förderung der numerischen und nichtnumerischen Angewandten Mathematik im Rahmen der Kollegialleitung des Mathematischen Seminars nicht effektiv genug nachkommen zu können, erkämpfte er alsbald die Umwandlung der Arbeitsstelle in ein eigenständiges Institut für Angewandte Mathematik, das er bis 1970 und von 1972 bis 1974 leitete. De jure blieb er bis 1970 auch Direktor des Mathematischen Seminars, und natürlich war er auch über viele Jahre Direktor des bereits erwähnten, aus dem Institutsverband 1963 ausgesiedelten Rechenzentrums.

Die gewaltige Breite des mathematischen Wirkens von Lothar Collatz und die daraus folgende Anziehungskraft auf jüngere Wissenschaftler/innen führte zu etwa 40 Promotionen unter seiner Anleitung, und er war an nahezu 30 Habilitationsverfahren beteiligt. Viele seiner Schüler versahen und versehen heute Professuren im In- und Ausland.

Auch privat fühlte sich Lothar Collatz vielen seiner frühreren Schüler lebenslang verbunden. In besonderer Weise galt seine Zuneigung wohl den Studenten aus jener Kriegsteilnehmergeneration, die 1945/46 zunächst in harter körperlicher Arbeit in Hannover Hörsäle und Seminarräume wieder soweit herrichteten, daß – ohne Heizung und eingewickelt in alte Militärmäntel, Handschuhe usw. – Vorlesungen durchgeführt werden konnten. Obschon nur etwa 10–15 Jahre älter als die Heimkehrer, ehemaligen Flak-Helfer usw., wurde er doch für diese z. T. entwurzelten, in den moralischen, geistigen und materiellen Zusammenbruch und in eine Trümmerwelt hineingestellten jungen Leute zu einer Leitfigur, die ihnen durch die voll eigener Begeisterung vermittelten klaren Prinzipien mathematischer Wissenschaft und wissenschaftlichen Disputs Halt verlieh und neue Ziele wies. Er selbst bezeichnete 1967 in einer Ansprache anläßlich der Trauerfeier für Hans Ehrmann, Jahrgang 1922 und Ordinarius an der TU Clausthal, diesen als seinen Lieblingsschüler. Wenngleich diese Hervorhebung einer Person wohl eher dem äußeren Anlaß zugeschrieben werden muß und noch viele Schüler später hinzukamen, beleuchtet sie doch diese besondere Verbundenheit, die auch durch die enge Freundschaft mit Julius Albrecht und anderen Vertretern dieser Jahrgänge bezeugt wird.

Wie viele herausragende Persönlichkeiten in Wissenschaft, Politik oder Kunst verstand es auch Lothar Collatz, seinen auf außerordentlicher Leistung beruhenden Bekanntheitsgrad durch außerberuflich-humorige „Markenzeichen" zu ergänzen und sich dabei auch fachfremden Personen unvergeßlich zu machen. Sportlich trainiert und behende, geschahen ihm z. B. merkwürdige Mißgeschicke: Im Anschluß an ein Essen zu seinem 75. Geburtstag wollte er Dankesworte sprechen und warf zunächst auf spektakuläre, wie eine Ungeschicklichkeit wirkende Weise das Mikrophon um, was den überraschten, aber für genaue Beobachtungsgabe bekannten Präsidenten der Universität Hamburg – einen Verehrer von Lothar Collatz – spontan zu dem bewundernden Ausruf „Was für ein Schauspieler!" veranlaßte. Ebenso ist der Tritt auf den langen Rock jener Solistin, die nach einem Konzert anläßlich der GAMM-Tagung 1966 in Darmstadt soeben von der Bühne abging, der Collatz in seiner Eigenschaft als GAMM-Vorsitzender jedoch im gleichen Augenblick einen Blumenstrauß zu überreichen gedachte, unvergessen. In diesen Zusammenhang gehören auch die von ihm mit seinen Studenten oder den Teilnehmern seiner Tagungen und deren Begleitpersonen durchgeführten Wanderungen, bei denen Wegweiser und angelegte Wanderwege bedeutungslos wurden zugunsten einer ausschließlich mittels Höhenmesser, Kompaß und Karte festgelegten Marschroute, die folglich auf Unterholz, Geröll oder Windbruch keinerlei Rücksicht nahm. Das Überqueren kleiner Wasserläufe abseits irgendwelcher Stege wurde gelegentlich eingeplant. Teilnehmenden Damen mit unpassendem Schuhwerk bot er dann zum Sprunge die hilfreiche Hand, wobei es vorkommen konnte, daß just auf diese Weise eine Landung im Wasser unvermeidlich wurde. Doch sein jungenhaftes Lachen und tausend schalkhaft-reuelose, mit seiner relativ hohen Stimme vorgebrachte Entschuldigungen ließen Unmut beim Opfer nicht wirklich aufkommen.

Sein Fotografieren zur Gewinnung eines ersten und immer auch eines zweiten Gruppenfotos anläßlich einer Tagung oder eines Abends in seinem gastlichen Hause geriet zu einer stets bemerkenswert komplizierten und dadurch international bekannten, von allen Beteiligten liebevoll tolerierten Zeremonie; unnötig zu erwähnen, daß natürlich alle Aufgenommenen mit Abzügen beschenkt wurden und so heute auch durch diese Bilder an viele schöne Begegnungen mit Lothar Collatz erinnert werden.

Die Mathematik in aller Welt und auch diese Zeitschrift, deren Gesicht Lothar Collatz über fünfunddreißig Jahre hinweg mitgeprägt hat, verlieren in dem Verstorbenen einen ihrer angesehensten Repräsentanten.

Literatur

1 COLLATZ, L.: Gleichungen geometrischer Ornamente. Zeitschr. für Mathem. Unterr. **64** (1934), 165–169.

2 COLLATZ, L.: Über das Differenzenverfahren bei Anfangswertproblemen partieller Differentialgleichungen. ZAMM **16** (1936), 239–247.

3 COLLATZ, L.: Konvergenzbeweis und Fehlerabschätzung für das Differenzenverfahren bei Eigenwertproblemen gewöhnlicher Differentialgleichungen zweiter und vierter Ordnung. Deutsche Math. **2** (1937), 189–215.

4 COLLATZ, L.: Konvergenz des Differenzenverfahrens bei Eigenwertproblemen partieller Differentialgleichungen. Deutsche Math. **3** (1938), 200–212.

5 COLLATZ, L.: Zur Interpolationstheorie der reellen periodischen Funktionen (zus. mit W. QUADE). Sitz. Ber. Preuß. Akad. Berlin 1938, 48 S.

6 COLLATZ, L.: Eigenwertprobleme und ihre numerische Behandlung. Akad. Verlagsges., Leipzig 1945.

7 COLLATZ, L.: Aufgaben monotoner Art. Arch. Math. **3** (1952), 366–376.

8 COLLATZ, L.: Inclusion theorems for the minimal distance in rational Tschebyscheff approximation with several variables. In: GARABEDIAN, H. L. (Ed.): Approximation of functions. Elsevier, Amsterdam-London-New York 1965.

9 BERTRAM, G.: Lothar Collatz zum 75. Geburtstag. Mitt. der GAMM, Heft 2 (1985), 14–17.

10 WERNER, B.: Einblicke in die frühen Arbeiten von Lothar Collatz. Bericht Nr. 6/85 des Inst. für Angew. Math. der Univ. Hamburg (1985), 12 S.

Hamburg RAINER ANSORGE

Zeitschrift für Angewandte Mathematik und Mechanik
Volume 71 1991 Number 7/8

Contents

Zeitschrift für Angewandte Mathematik und Mechanik is indexed in Current Contents/Physical, Chemical & Earth Sciences; Science Citation Index; Mechanics; Applied Mechanics Reviews; Mathematical Reviews; Zbl. Math.

Bei Fragen zur Produktsicherheit wenden Sie sich bitte an:
If you have any questions regarding product safety,
please contact:

Walter de Gruyter GmbH
Genthiner Straße 13
10785 Berlin
productsafety@degruyterbrill.com